DISCARD

5-06

Sandy Berger's
Great Age Guide to Gadgets and Gizmos

Contents at a Glance

800 East 96th Street,
Indianapolis, Indiana 46240 USA

Sandy Berger's Great Age Guide to Gadgets and Gizmos

Copyright © 2006 by Que Publishing

International Standard Book Number: 0-7897-3441-9

Library of Congress Catalog Card Number: 2005922885

Printed in the United States of America

First Printing: September 2005

08   07   06   05   4   3   2   1

Trademarks

All terms mentioned in this book that are known to be trademarks or service marks have been appropriately capitalized. Que Publishing cannot attest to the accuracy of this information. Use of a term in this book should not be regarded as affecting the validity of any trademark or service mark.

Warning and Disclaimer

Bulk Sales

Que Publishing offers excellent discounts on this book when ordered in quantity for bulk purchases or special sales. For more information, please contact

> U.S. Corporate and Government Sales
> 1-800-382-3419
> corpsales@pearsontechgroup.com

For sales outside of the U.S., please contact

> International Sales
> international@pearsoned.com

ASSOCIATE PUBLISHER
Greg Wiegand

ACQUISITIONS EDITOR
Stephanie J. McComb

DEVELOPMENT EDITOR
Laura Norman

MANAGING EDITOR
Charlotte Clapp

PROJECT EDITOR
Tonya Simpson

COPY EDITOR
Karen A. Gill

INDEXER
Chris Barrick

PROOFREADER
Susan Eldridge

TECHNICAL EDITOR
Dana Carroll

PUBLISHING COORDINATOR
Sharry Lee Gregory

BOOK DESIGNER
Anne Jones

PAGE LAYOUT
Nonie Ratcliff

GRAPHICS
Tammy Graham

Table of Contents

About the Author

Sandy Berger, nationally respected computer authority, journalist, media guest, speaker, and author, has more than three decades of experience as a computer and technology expert.

As president of Computer Living Corp, a computer consulting and training company, Sandy applies her unique ability to explain in easy-to-understand language how to use and enjoy today's technology. Her trademark, "Compu-KISS®" which stands for "The Computer World—Keeping It Short and Simple," represents her approach to helping others enhance life through the use of computers and technology.

As primary content provider and host of AARP's Computers and Technology website, Sandy has her finger on the pulse of the boomer and zoomer community. Her feature stories, product reviews, and computer tips have brought a special insight and ease of use to millions of boomers and zoomers.

Sandy writes a monthly column in *Smart Computing* magazine called TechMates, which reviews two high-tech products that complement each other. She has been a guest on hundreds of radio and television shows, including NBC's TODAY Show, NBC News, CBS News, FOX News, ABC News, WGN, and WOR radio.

Sandy is an excellent example of her own philosophy—use technology, but keep it short and simple. Her previous three books, *How to Have a Meaningful Relationship with Your Computer*, *Your Official Grown-Up's Guide to AOL and the Internet*, and *Cyber Savers—Tips & Tricks for Today's Drowning Computer Users*, adhere to these principles.

Sandy is a consumer advocate who promotes simplicity, ease of use, and stability in consumer technology products. She works with hardware and software developers to help them make their products more user friendly.

A cum laude graduate of Chicago's DePaul University, Sandy went on to complete intensive IBM training in computer systems, analysis, programming, system operations, and numerous computer languages. She subsequently applied her expertise within several major corporations before founding Computer Living Corp.

My fondest wish is that you enjoy this book and have fun with technology!

Sandy Berger

Dedication

In loving memory of my father, LeRoy Truschke, who taught me to enjoy the little things in life. He always loved gadgets and gizmos. Dad, this one's for you!

Acknowledgments

When I was growing up, we spent the summers in a small town outside of Chicago. Each year, the Fourth of July festivities ended with a tug-of-war between the men of the community. The winners celebrated their victory for the rest of the summer.

Creating a book like this is a little like a tug-of-war. It can be a constant struggle to get everything done in a timely manner. Now that the book is completed, I would like to celebrate with and to thank all of those who were on my team, constantly helping me pull that rope.

Thanks to Paul Boger, Greg Wiegand, Stephanie McComb, Sharry Lee Gregory, Laura Norman, Anne Jones, Judy Taylor, Karen Gill, Dana Carroll, Greg Yurchuck, Lisa Jacobson-Brown, Marybeth Sandell, and the entire staff at Que.

Special thanks to my husband, Dave, who is always helping me tug that rope. Also thanks to the companies who created and provided these great gadgets and gizmos.

I celebrate all of you. Thanks to all!

We Want to Hear from You!

As the reader of this book, *you* are our most important critic and commentator. We value your opinion and want to know what we're doing right, what we could do better, what areas you'd like to see us publish in, and any other words of wisdom you're willing to pass our way.

As an associate publisher for Que Publishing, I welcome your comments. You can email or write me directly to let me know what you did or didn't like about this book—as well as what we can do to make our books better.

Please note that I cannot help you with technical problems related to the topic of this book. We do have a User Services group, however, where I will forward specific technical questions related to the book.

When you write, please be sure to include this book's title and author as well as your name, email address, and phone number. I will carefully review your comments and share them with the author and editors who worked on the book.

Email: feedback@quepublishing.com

Mail: Greg Wiegand
Associate Publisher
Que Publishing
800 East 96th Street
Indianapolis, IN 46240 USA

For more information about this book or another Que Publishing title, visit our website at www.quepublishing.com. Type the ISBN (excluding hyphens) or the title of a book in the Search field to find the page you're looking for.

Foreword

Dear Great Age reader,

I've always had a fascination with innovations that affect the way people get things done. In fact, you could say that this pursuit has defined a very important aspect of my life.

I was given a gift of a good education, in Australia. My family experiences taught me the importance of working hard and dreaming big. Through a series of fortuitous events, that I describe in my book, Tom Crow—*King of Clubs*, I landed in the wonderful game of golf. By 1961, I had won The Australian Amateur Championship (AAC).

During this time, I grew to love playing the game and to love the fellowship that is so much a part of the sport. I took up an opportunity on the business side of the game, representing an Australian club manufacturer. I began to think about various technologies and equipment design enhancements that could improve the golf experience. Then, in 1973, I made the big leap. My wife Cally and I moved from Australia to California with our young family. With the help of many good people and a $60,000 loan, we started a business called *Cobra Golf*.

Two decades later, our company had launched products like the Baffler, the three-wedge system, the long-shafted driver (dubbed the "Long Tom"), and a host of other industry-leading initiatives. These came to fruition just as golf was emerging as a major industry, enabling us to sell that business for $750 million dollars.

It was an incredible journey. Great Age guide Sandy Berger guessed it right when she talks not just about the latest and greatest innovations, but about the way that effective gadgets and gizmos can take your game to a whole new level. I'm talking now about the game of life. Technology can be hard, even frustrating; but if the goal is living a rewarding life and all that is involved (relationships, challenges, communication, travel, time management, health, finances—yes, even good golf) then technology will come into play.

Bobby Jones liked to say that "Golf is the closest game to the game we call life". As in golf, so it is in life. If you want to know which technology is best and what it means for you, this book will help you get the answers.

Wishing you the best,

Tom Crow
Founder, Cobra Golf
Co-Executive Producer, Bobby Jones, Stroke of Genius

Have you ever stopped to think about the role of technology in your life? There's no doubt that today's younger generations will benefit from current and future technological advances, but today's older generations are already seeing the biggest lifestyle improvements ever. We are living longer, more active lives than our parents and grandparents. Our ancestors went from youth to middle age to old age. We, with a new mentality and the help of technology, have added an entire epoch to our lives—the Great Age!

Everyone knows that the term *baby boomer* refers to individuals who were born after World War II. This group is generally recognized as encompassing people who were born between 1946 and 1964. The boomers' older siblings don't have a moniker associated with their generation, but along with the boomers, they are often referred to as *zoomers* because they are not ready to be relegated to a rocking chair. Boomers and zoomers are zooming into the latter part of their lives, zooming into technology, and zooming into everything they do. They are vibrant individuals who deal enthusiastically with all aspects of their lives. They are ready to enjoy the Great Age that they have created.

About *Sandy Berger's Great Age Series*

It's about time that someone addressed the issues that face those of us who did not grow up with computers. We are not technologically impaired. We are not dummies. And we are not about to be over-looked.

It's just that we didn't learn about computers in school, so we sometimes approach the new-fangled digital world with a bit of trepidation. Can someone please tell us just what we need to know without the complicated mumbo-jumbo?

That is exactly what this series does. It is explicitly geared for the needs and wants of baby boomers and beyond. It tells you just what you need to know—no more and no less. It uses the winning formula of need-to-know information along with easy-to-understand explanations.

Over the past decade, I have helped many boomers and their older siblings learn how to use computers and technology to enhance and improve their lives. I understand the needs and wants of this generation. After all, I am enjoying the Great Age myself. I am anxious to help guide you into the world of technology where the Great Age is full of enjoyment and anticipation.

Gadgets and Gizmos

These are not your grandfather's gadgets. Neither are they your grand-children's gadgets. These are chosen just for you. As we age, we realize that the path to better living is through technology. Computers and technology now affect every aspect of our lives. They have extended our longevity, improved our health, made travel easier, and even made it easier to have fun.

Yet technology is moving so quickly that it is difficult for the average person to keep up with all the useful products that currently are available. Did you know that there is a nondrug product to lower blood pressure? A mouse to alleviate hand and wrist strain? A light to help you see more clearly? A camera for shaky hands? A wheelchair that climbs stairs? All these products are discussed in this book. I've scoured the planet to fill *Sandy Berger's Great Age Guide to Gadgets and Gizmos* with unique products that are useful and fun. These gadgets and gizmos put you on the cutting edge while making sure you are far from the bleeding edge.

You can use this book as a purchasing guide, but you don't have to. You can simply use it to learn about the many gadgets and gizmos that are currently available to you. From these products, you also get a glimpse of what technology to expect in the next few years. This book is intended to give you a taste of the richness and ease that technology can bring to your life. You'll get an idea of the many products that can enhance your life.

Remember that you don't have to memorize anything here. Just sit back, relax, and absorb whatever information is important to you. This book gives you the gist of what today's gadgets and gizmos can do for you. After you know what is out there, you can experience as much or as little as you like. You're under no pressure. Through this book you will come to enjoy technology.

What's Inside

This book comes with no special instructions for using it. Start at the beginning, or jump around as you please. I've added a quote at the beginning of each chapter because we can always learn from others. I've also included a few special features to help you in your quest for knowledge.

Items that get my official seal of approval are marked as Sandy's Favorite. These are useful, cost-effective products, many of which I use every day.

At the end of each chapter is a list of terminology. After all, part of dealing with technology is learning the lingo.

How This Book Is Organized

There are gadgets and gizmos for every aspect of our lives. I've gathered a wide varity of devices for you to consider, which are grouped according to their applications.

- Chapter 1, "Keyboards, Mice, and More," offers a wide variety of keyboards, mice, and other input devices.

- Chapter 2, "Computer and Internet Gadgets," is filled with fun and useful gadgets to use with your computer.

- Chapter 3, "Entertainment Gadgets," brings the digital world of audio, music, and video into your life.

- Chapter 4, "Fun and Games," lets you know that you are never too old to enjoy high-tech gadgets made just for fun.

- Chapter 5, "Vision Aides," helps you protect and enhance your vision as you age.

- Chapter 6, "Accessories for Healthy Living," can help you stay fit and healthy by using high-tech gadgets.

- Chapter 7, "Communication Devices," helps you keep in touch with family and friends.

- Chapter 8, "Gadgets for the Home," makes living at home easier and more fun.

- Chapter 9, "Photography and Printing," investigates products that enable you to capture and print those precious memories.

- Chapter 10, "Travel and Automotive," makes your life on the road easier than ever before.

Chapter 1

Keyboards, Mice, and More

If you build a better mousetrap, you will catch better mice.
— George Gobel

Computers have changed dramatically over the past 10 years, but the keyboards and mice that are included with most of today's computers are the same. Better alternatives are available, and this chapter shows you many of them. Keyboards can be easier to see and more comfortable to use. Mice and other input devices come in many different shapes and sizes to accommodate a large variety of different users. There are even devices for someone who has shaky hands, those made for use with one hand, and those made for arthritic hands. When it comes to computer input devices, there truly is something for everyone.

  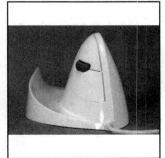 

Ergonomic Keyboards

Whether you are a trained touch typist or you hunt and peck at the keyboard, the computer keyboard should be an asset to your productivity. You don't have to use the keyboard that came with your computer. Many different types are available.

Typewriters were first introduced iin the late 1800s. Back then, everyone was so excited about a machine that could produce writing that no one ever considered how the typewriter would affect the typist. While typing, the hands are held in an unnatural position. Touch typists who spent long hours at the keyboard were plagued by hand and wrist injuries, in addition to stress and tension in the back and neck.

The computer keyboards that we use today are based on those archaic typewriters. The same key placement and structure was kept to accommodate millions of people who had already learned to type. So the same stresses and injuries now occur with computer keyboarding. Because more and more people use a computer every day, these problems have become widespread.

Ergonomic keyboards were created to help relieve the stress on wrists, shoulders, and arms. These keyboards use various methods to place the hands in a more natural position. Some keyboards adjust the keys only minimally, whereas others move the keys into completely new configurations. Almost all provide a more natural typing position. Although ergonomic keyboards are available for everyone, they are especially soothing for people who have done a lot of typing in their lives and those who have dexterity problems with their fingers.

Kensington Comfort Type Keyboard

This is a regular computer keyboard except that the keys are slightly angled to keep the hands and wrists in a more comfortable position. Most people don't need to

Model: Comfort Type USB/PS2
Manufacturer: Kensington
URL: www.kensington.com
Price: $19.99

alter the way they type to use it. The Comfort Type keyboard allows your hands to rotate slightly so that the index fingers are angled toward each other instead of being parallel to each other. It doesn't seem like a big change, but remember that when it comes to ergonomics, even small adjustments can help eliminate body stress and strain.

SafeType Keyboard

Yikes! This is one wacky-looking keyboard. It consists of three joined parts. The middle lies flat on the desk. The two side portions rest vertically next to the middle section with the keys on the outside. You position your hands up and down in a hand-shake position on the outside of the keyboard and type. Two small mirrors help you situate your hands. SafeType is designed to keep your hands in their most natural position just as you would hold them when walking. This keyboard is for touch typists only. It takes some getting used to, but it can be "just what the doctor ordered" for wrist problems that are adversely affected by the use of a regular computer keyboard.

Model: SafeType
Manufacturer: SafeType
URL: **www.safetype.com**
Price: $159

Kinesis Contoured Keyboard

This Kinesis contoured keyboard is another unusual-looking keyboard that is ergonomically designed. The keys are arranged in two concave "wells," which are separated from each other by several inches. Each hand fits comfortably in its own area. The keyboard actually forces the hands and arms to be in a better position than regular keyboards. The letter and number keys are in the same positions that they maintain on a regular keyboard. However, because each "well" is rounded, the relative spacing of the keys is slightly different from a regular keyboard and takes some getting used to.

Model: Contoured Keyboard
Manufacturer: Kinesis
URL: **www.kinesis-ergo.com/**
Price: $159

Big Key Keyboards

Using a computer can pose problems for aging eyes. If you strain your eyes to see those tiny letters on the keyboard, I have the answer for you. Now some keyboards have larger keys, or you can place inexpensive stickers on your current keyboard. Try these, and you'll soon be singing, "I can see clearly now!"

BigKeys LX Keyboard with USB

This keyboard has one noticeable feature: The keys are really, really big. In fact, each key is one-inch square, which is about four times larger than the keys on a standard keyboard. BigKeys is available with black letters on bright white keys, multicolor keys, or bright white letters on black keys. It is also available with the keys laid out in QWERTY order like most keyboards, or in ABC order. BigKeys also has an "assist mode" to aid keyboarders who find it difficult to press multiple keys at the same time. In

addition, this keyboard prevents run-ons, or the effect of a single character being repeated if you press the key for too long. Depressing a single key causes

> **Model:** BigKeys LX Keyboard with USB
> **Manufacturer:** Greystone Digital, Inc.
> **URL: www.bigkeys.com**
> **Price: $167**

only one character to be sent to the computer, regardless of how long it is held down. Because the keys are larger than normal, touch typists might have difficulty with this keyboard, but computer users who require large keys so that they can locate and operate a keyboard will love it.

VisiKey Keyboard

The full name of the new VisiKey keyboard is the Enhanced Visibility Internet Keyboard. Even a quick glance at the keyboard will make you notice its main feature: The letters, numbers, and symbols on the keys are bright white, and they are much bigger and more visible than those on regular computer keyboards. In fact, the letters are 430% larger than the letters on a standard keyboard. Unlike the BigKey keyboard, the keys on the VisiKey keyboard are normal sized and well-suited to touch typing. Only the letters are larger. This enhanced visibility lettering system gets a rating of 20/300 on the Snelling visual acuity scale. Compare that to most standard keyboards, which are rated 20/70, and you can get a feel for the visual ease that the VisiKey keyboard provides.

Model: VisiKey Keyboard
Manufacturer: VisiKey
URL: www.visikey.net
Price: $39.95

VisiKey has all the normal computer keyboard keys plus 15 extra hotkeys. The top of the keyboard has media center controls for playing, pausing, muting, and so on for when working with music and video files with your media player. A separate area next to the media center has Internet keys that give one-touch email, web search, browsing, and forward and back keys. The arrow navigation keys are curved, and this ergonomic shape makes them available by touch without looking at the keyboard.

20/20 Type—Large Print Keyboard Labels

If you can't see the letters on your keyboard, but you don't want to purchase a whole new keyboard, you can attach these peel-and-stick vinyl labels to the keyboard you already own. The print is four times larger than the print used on standard keyboards. Choose from black letters on a white background or white letters on a black background. Each set includes all the alphanumeric characters plus the special keys on the computer keyboard such as Escape, F1, Alt, and Ctrl. The labels attach firmly but will eventually start to peel with heavy use.

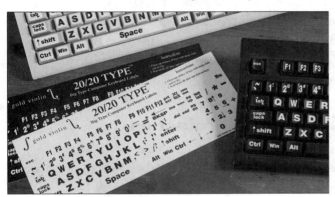

Model: Gold Violin Large Print Keyboard Labels
Distributor: Gold Violin
URL: www.goldviolin.com
Price: $8

Specialty Keyboards

If you aren't motivated to give up your plain vanilla keyboard for the practical reasons like comfort or ease of visibility, there are many other inventive keyboards that do everything from keeping you organized to allowing you to use your fingerprint as a username and password that you'll want to check out. I bet you'll be surprised by some of the creative ideas that have been incorporated into computer keyboards.

Cordless Keyboard with Microban Protection

You don't have to be a verminophobe to have considered how many germs reside in a computer keyboard and mouse. When family members sneeze on a keyboard, are they spreading dangerous germs? If you have ever asked yourself this question, this antimicrobial keyboard is sure to give you some peace of mind. The built-in antimicrobial protection inhibits the growth of bacteria and other microbes. Fellowes says this special protection also makes the keyboard last longer. The Microban protection is built into several other Fellowes products. You can also get an ergonomic split design keyboard, a cordless keyboard, several mice, and mouse pads with the same germ protection.

Model: Cordless Keyboard with Microban Protection Item #98916
Manufacturer: Fellowes
URL: **www.fellowes.com**
Price: $59.99

Optical Desktop with Fingerprint Reader

If you are tired of trying to remember your username and password for each of the websites you visit, this Microsoft keyboard might be the one for you. It has a built-in fingerprint reader that allows you to log in to websites with the touch of a finger. The fingerprint reader identifies you, and the software enters your username and password. The optical desktop combines this neat keyboard with an excellent Microsoft optical mouse. Both are wireless, offering even more functionality. The keyboard also has multimedia keys for easy one-touch access to volume, music, and video controls.

Model: Optical Desktop with Fingerprint Reader
Manufacturer: Microsoft
URL: www.microsoft.com
Price: $84.95

Keyboard Organizer

The MKO-1900 is a fully functioning computer keyboard with a surprise inside. The keyboard is hinged on the back. Lifting up the front reveals compartments to keep your desktop gear organized. There is an area that's made for storing CDs or DVDs. Several other compartments can be used for business cards, pens, papers,

Model: MKO-1900
Manufacturer: MyKeyO
URL: www.keyboardorganizer.com
Price: $24.95

stamps, paperclips, or other miscellaneous items. Special hinges on the keyboard allow it to stay open until you want to close it. There is also a backlit version that lights the keys with a cool blue glow at the touch of a button. MKO-1900 is even available wireless.

FrogPad

This one-handed keyboard is perfect for the physically challenged or for those needing mobile computing. This sturdy little device has 15 main number/letter/symbol keys, surrounded by several toggle keys. Unlike the QWERTY layout that is used on most computer keyboards, these keys are laid out for functionality, with the most-used keys right under your resting fingers. You'll definitely experience a learning curve with this device because it is completely different from anything you've ever seen before. However, because the FrogPad is so intuitive, learning to use it is easier than learning to touch type on a regular keyboard. Unlike more cumbersome methods of one-handed typing, you only need two fingers to type any

Model: FrogPad USB
Manufacturer: FrogPad USB
URL: www.frogpad.com
Price: $169.99

letter on the FrogPad. It comes with a good instruction manual and an excellent Quick Start Guide. You need to choose between the left-hand or right-hand model. It uses a USB cable to hook up to any computer. A Bluetooth version is also available for use with a Bluetooth-compatible cell phone or PDA.

Logitech diNovo Media Desktop

Sandy's favorite

If you've ever tried to lean back and use the keyboard in your lap, you probably found yourself tethered to the computer by a restrictive wire. That's why wireless input devices have become one of today's biggest trends.

While many wireless keyboards and mice use older technologies to make the cordless connections, a newer wireless technology called Bluetooth has proved to be more reliable for cordless devices.

The Logitech diNovo Media Desktop combines a Bluetooth keyboard, mouse, and numeric pad. The extra keypad is a feature that makes the diNovo stand out from the competition. It is a numeric keypad, but it has several other useful features, too. It displays the date, time, and temperature and doubles as a mini-calculator. The numeric pad also works as a control pad for all your digital media. It features one-button access to music, photos, and email.

Model: Logitech diNovo Media Desktop
Manufacturer: Logitech
URL: www.logitech.com
Price: $249.95

The keyboard, like most other wireless keyboards, uses regular batteries that you have to replace occasionally. The mouse, however, has a welcome rechargeable battery. The Bluetooth receiver conveniently doubles as the mouse recharging station. The mouse includes a scroll wheel and programmable buttons. To use this Bluetooth device, your computer must be equipped with the Bluetooth technology.

The diNovo Desktop also comes with a travel case for the mouse and numeric pad. Included is a small USB mini-receiver that can plug directly into your laptop. You can use this input system on the road.

The diNovo is a bit pricey, but its great tactile feel and numerous features make it a worthwhile addition to any computer system.

Keyboard Accessories

A good keyboard is a must for keeping your hands and wrists in tip-top condition. Many accessories for the keyboard can also make your computing just a little easier. There are some computer accessories that I wouldn't do without. A keyboard shelf, like the Fellowes Adjustable Keyboard Manager, puts the keyboard and mouse at the proper height to alleviate stress on the wrists. I use one at work and at home. You might also want to try a gel wrist pad for comfort when you are using the computer. And don't forget the tools you need to keep the keyboard clean. Didn't your mother tell you that cleanliness is always worthwhile?

Fully Adjustable Keyboard Manager

Did you know that having your keyboard and mouse on the average height desk puts it too high for comfortable computing? The keyboard and mouse should be several inches below desk level to keep your hands, arms, and wrists more comfortable. If you have an open desk or table, this keyboard manager puts the mouse and keyboard at just the right level. The keyboard tray attaches to the underside of your desk. It comes with a mouse tray that can be mounted on either the left or right. A simple knob adjusts the height, tilt, and position of the keyboard.

Model: Fully Adjustable Keyboard Manager
Manufacturer: Fellowes
URL: **www.fellowes.com**
Price: $79.99

Gel Wrist Pillow

If you want a little extra support for your wrists, this nifty wrist pillow is for you. It is wider than most wrist supports, spanning the entire width of most keyboards. A nonskid base keeps the pad securely in place. The puncture-proof gel is soft and soothing.

Model: Gel Wrist Pillow
Manufacturer: Kensington
URL: www.kensington.com
Price: $15.99

AirDR Dust Blaster CO_2

You can clean your keyboard by turning it upside down and shaking it to dislodge any dust and food fragments, but even then, crumbs and dirt can get lodged in the crevices. It's good to give your keyboard and other equipment an occasional shot of air to get rid of unwanted dirt, dust, and foreign particles. AirDR Blaster uses environmentally safe, antistatic, non-flammable air cartridges. A special airflow control delivery system puts just the right amount of air pressure on your precious equipment. Use a full blast for pin-point clean-ing of electrical equipment or a gentle burst to clean off negatives before you scan them into your computer.

Model: AirDR Blaster CO_2
Manufacturer: Digital Innovations
URL: www.digitalinnovations.com
Price: $12.99

Data-Vac Shuttle Vacuum

If you feel more comfortable sucking the dust and dirt out of your computer equipment rather than blowing air at it, try this shuttle vacuum. It is more powerful than many others. It comes with its own stand, which doubles as a recharging station for the four AA rechargeable batteries that are also included. The vacuum includes a tiny dust brush, a crevice tool, and a blower adapter. Use the rechargeable batteries or power it with regular batteries or the included power adapter. It's a little pricey if you plan to use it just to clean the keyboard, but it might be worthwhile if you have other equipment that you want to keep dust-free.

Model: Data-Vac Shuttle Vacuum
Manufacturer: Metropolitan Vacuum Cleaner
URL: www.metrovacworld.com
Price: $27.99

Ergonomic and Assistive Mice

Unlike the keyboard, the computer mouse has evolved considerably since its inception. The original mouse had only one button, but today's mice usually have between two and five buttons that can be programmed to perform various functions. A wheel often appears between the two main buttons that allows users to scroll up and down web pages and documents. Mechanical mice that have a rubber or metal ball on the bottom have all but been replaced by the more precise optical mouse that uses a red-glowing laser to detect the mouse's movement.

Yet even with all this new functionality, the mouse is not suited to everyone. People who have large hands sometimes find their mouse too small and the buttons difficult to press. Arthritic hands often have trouble pressing the keys on the mouse. Shaky hands need something more substantial to hold on to. Don't despair! You don't have to use the mouse that came with your computer. You can use several alternative devices instead of a mouse, or one of many alternative types of mice. One of them is sure to be perfect for you!

An optical mouse uses a light-emitting diode, an optical sensor, and a digital signal processor instead of a mechanical ball to detect the movement of the mouse. You can identify an optical mouse by the light that it emits from its bottom. Optical mice are precision instruments that don't need a mouse pad.

AirO$_2$bic Mouse

Sandy's favorite

This mouse has an unusual shape. It has a large elliptical bump next to a scooped-out area where you rest your hand. It is called a gripless mouse because you don't have to hold the mouse in any way. You simply rest your hand comfortably in the curved area and glide the mouse across the desktop. Because the AirO$_2$bic mouse is larger than a regular mouse, it seems cumbersome at first. However, it

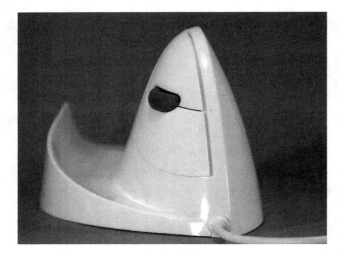

Model: The Virtually HandsFree
System
Manufacturer: Designer Appliances,
Inc.
URL: **www.quillmouse.com**
Price: $149.95

works seamlessly and is easy to get used to. The mouse can be purchased separately for $99, but it really shines when you purchase the entire system, which includes the clickless software. With this software, you never have to press a button. The navigation is easy and the system ingenious. This is the perfect input device for those with Parkinson's, arthritis, and other conditions that result in a loss of physical dexterity. The $AirO_2bic$ mouse has even garnered an ease-of-use commendation by the Arthritis Foundation. And you can use it in either hand.

Evoluent VerticalMouse 2

I've seen a lot of different input devices, but this is the first one I've seen that is purple and black. Don't let the colors turn you off. The VerticalMouse is a serious ergonomic tool. It has a footprint that is about the same size as a regular mouse, but it is about 4 times as tall. It looks like a large purple and black mound. You rest your hand on top of the mound in a handshake position. The grip is similar to an ordinary mouse, but your hand is turned sideways and is in a more comfortable

position. Your hand rests on the black portion of the mouse, which has a soft, comfortable feel. The mouse is nicely molded to fit the shape of a

Model: Evoluent VerticalMouse 2
Manufacturer: Evoluent
URL: www.evoluent.com
Price: $74.95

hand. Because you can rest your entire hand on the mouse, it helps steady any shakiness you might have, making it a good choice for many. The main buttons are on the outside of the mouse, so if you are left handed, be sure to purchase the left-hand model. The mouse has four programmable buttons and a scroll wheel and is made for average to large hands. If you have a small hand like mine, you might find the scroll wheel a bit hard to reach. Undoubtedly, this mouse is more comfortable than the average mouse.

RollerMousePRO

Sandy's favorite

Using a mouse sometimes causes wrist and hand strain. The repetitive motion of moving the hand between the keyboard and the mouse can also cause shoulder strain. The RollerMouse solves these problems by putting the mouse functionality right below the keyboard. You don't have to move your hand over to the mouse to click or scroll. The RollerMouse is attached to a platform for the keyboard. Just position the keyboard on the platform, plug in the USB adapter, and it is ready to go. A 7-inch roller bar that looks like a large black rubber straw controls the cursor movements on the screen. Pressing the bar is like left-clicking the mouse. RollerMouse has a scroll wheel and five buttons that you can set in three different modes. This mouse takes some getting used to, but it is a great product for anyone who uses the computer a lot.

Model: RollerMouse PRO
Manufacturer: Contour Design
URL: www.contourdesign.com
Price: $199

Assistive Mouse Adapter

Millions of people around the world suffer from Parkinson's disease and other conditions that cause hand tremors. These involuntary movements make it difficult for hand sufferers to use the standard computer mouse. The solution is the Assistive Mouse Adapter, which was developed by IBM. This adapter is a small rectangular device that is plugged in between the computer and the mouse. Designed to work with any PC, the adapter functions like the stabilization in some camcorders to filter out the shaky movements. The Assistive Mouse Adapter compensates for shaky movements and has a sensitivity setting that can be adjusted according to tremor severity. It can also eliminate the multiple clicks that a shaky hand might produce.

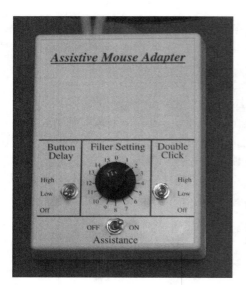

Model: Assistive Mouse Adapter
Manufacturer: Montrose Secam Limited
URL: www.montrosesecam.com
Price: $99

Wireless IntelliMouse Explorer

This is the mouse that comes in the Optical Desktop with Fingerprint Reader package mentioned in the previous "Specialty Keyboard" section of this chapter. IntelliMouse is worth looking at alone because it has a comfortable ergonomic design that is good for most hands. As with most wheel mice, you can use the scroll wheel for vertical scrolling. This mouse, however, has a special tilt wheel that allows you to scroll horizontally on Web pages and documents, besides vertically. The wireless receiver works quite well. You simply plug it into a USB port, and you can

Model: Wireless IntelliMouse Explorer
Manufacturer: Microsoft
URL: www.microsoft.com
Price: $44.95

move the mouse without being tethered to the computer. Long-lasting batteries and optical precision make this a smart choice in the more regular mouse category.

Gyration Ultra

This input device will really have you mousing around. You can use it on your desktop like a regular mouse, or you can hold and move it in midair to move the cursor on the screen. The Gyration's motion-sensing technology translates your movements, and a miniature gyroscope keeps everything balanced. The wireless technology performs well, and the Gyration is surprisingly accurate. If you sometimes feel the stress in your wrist or arm from using a regular mouse, you can give your wrist a rest by using the Gyration in the air. There is virtually no learning curve. It also comes with software that lets you move forward and back on Web pages with the flick of your wrist. This is an excellent input device for presentations, web surfing, and gaming.

Model: Ultra GT Cordless Optical Mouse
Manufacturer: Gyration
URL: www.gyration.com
Price: $79.99

Mouse Alternatives

What if you don't want to use a mouse at all? Well, you don't have to. Several alternatives to the common computer mouse are available. These include trackballs and touch pads. These devices have all the functionality of a mouse, and some have additional features that make them even easier to use. Physical problems might lead you to one of these alternative mice, but many people use them just because they find them simpler and easier than using a mouse.

Orbit Optical Trackball

The Orbit Trackball stays in one place on your desktop. To move the cursor on the screen, use any finger or even the palm of your hand to move the large ball on the top of the trackball. Two large buttons on either side of the ball perform the same functions as the left and right mouse buttons. The optical tracking makes this as precise as any optical mouse. Many people use it because it is easier to use than the normal mouse, but it also comes in handy if you have large hands or physical dexterity problems. You can use it with either the right or the left hand.

Model: Orbit Optical Trackball
Manufacturer: Kensington
URL: www.kensington.com
Price: $29.99

Cirque Smart Cat

This is yet another input device that is a completely different type of mouse substitute. It is similar to the touch pads that come on some laptop computers, but it is larger and easier to use. The Smart Cat has all the functionality of a mouse. Often, it is easier to perform mouse tasks using the Cat. For instance, gliding a finger

Model: Smart Cat Touch Pad USB
Manufacturer: Cirque
URL: www.cirque.com
Price: $64.95

along the edge of the pad activates vertical and horizontal scrolling in addition to magnification. Tapping the upper-right portion of the touch pad simulates right-button mouse clicking. The Smart Cat has four customizable buttons. The intuitive software allows you to customize the sensitivity to suit your needs. Each operation has a distinctive sound, so you always know exactly what you're doing. The Smart Cat works with PCs and Macs. All versions of Windows (Windows 98 and better) and Macs (MacOS 8.5 or better) are supported except for Windows Me.

TrackMan Wheel

Like the Orbit, this trackball sits stationary on the desktop. You move the large red ball on the side with your thumb. The TrackMan has three customizable buttons and a scroll wheel. The optical tracking makes the trackball respond well. Unfortunately, this version is for right-handed people only. Logitech also makes a MarbleMouse mouse with the ball on the top that you can use with either hand.

Model: TrackMan Wheel
Manufacturer: Logitech
URL: **www.logitech.com**
Price: **$29.95**

Sandy's Lingo List

The world of technology has created some crazy new words. Here are explanations for a few of the more unusual words used in this chapter.

Bluetooth—A short-range technology that uses radio frequencies to connect devices wirelessly. Bluetooth technology is embedded in a computer chip and can be used in many different devices. A Bluetooth camera can transmit its pictures to a Bluetooth computer or printer without being attached.

ergonomics—The science that studies the safety and comfort of machines and furniture for humans. Ergonomically designed furniture and computer equipment helps to avoid repetitive stress injury, back and neck strain, and other computer-related physical problems.

optical mouse—A mouse that uses a laser to detect movement. Optical mice are more precise than older mice that have a wheel and ball on the bottom. Optical mice have no mechanical moving parts. They don't require the use of a mouse pad and work anywhere except extremely shiny surfaces like glass and mirror.

QWERTY—The term used to describe the standard English-layout keyboard. Q, W, E, R, T, and Y are the first six letters on the keyboards below the row of numbers.

repetitive stress injury (RSI)—A type of injury where soft tissue in the body, such as muscles, nerves, tendons, and joints, becomes irritated or inflamed. If the injury goes untreated, permanent damage can occur. These injuries are known to be caused by repetitive motions. Musicians and others who perform the same physical motions over and over are often targets of RSI. When dealing with computers, the repeated motions and prolonged use of the keyboard or mouse can lead to RSI in the hands, wrists, elbows, and arms. Simple measures such as taking occasional breaks from computer work and ergonomically correct positioning of the keyboard and mouse can often prevent repetitive strain injury.

scroll wheels—Often found between the two buttons of a mouse, they can be used to scroll on web pages and documents.

touch pads—Small touch-sensitive pads that can be used in place of a mouse. You drag your finger across the pad to move the pointer on the screen. Tapping on the pad replicates the clicking of the mouse.

USB—Universal serial bus is a way of connecting external devices to a computer. USB allows data to be transferred between the connected device and the computer much faster than older serial or parallel connections. A USB device is easy to install. On most newer computers, you just plug the device into the USB port on the computer, and it works without additional software. All Macs and PCs sold in the past several years have USB ports.

Chapter 2

Computer and Internet Gadgets

Hardware: The parts of a computer system that can be kicked.

—Jeff Pesis

Computers and the Internet have revolutionized our lives. They have also sparked the production of a large number of gadgets and gizmos. Have you ever wandered through a computer store and not even recognized the uses of some of the displayed devices? Well, some of those products might be just what you need to help you get on the Internet, get organized, or protect your computer equipment. This chapter introduces you to just a few of the many computer-related devices that can make your life easier. These digital devices can help you create a wireless network, transport files, and even identify you by your fingerprint.

Hot Digital Devices

It's fun to play with the latest and greatest computer gadgets and gizmos. It's even more fun when these devices turn out to truly useful. I've chosen a few great products that are on the cutting edge but will keep you away from the bleeding edge.

SWISSMEMORY Knife

This little device is from Victorinox, the renowned Swiss manufacturer of the Swiss Army Knife that has been supplying knives to the Swiss army from more than 100 years. It looks just like a Swiss Army knife, but hidden away with the knife, scissors, file, and screwdriver is a fully functioning USB data storage device. Unfold the USB hard drive, plug it into your computer, and you see another hard drive listed in the My Computer area. You can copy documents, audio files, and pictures to your "knife" and carry them with you wherever you go. Of course, if you are flying, this device is available without the knife and other tools. The SWISSMEMORY device works with PCs and Macs and comes in sizes up to 1GB. This gizmo is amazing. It's small, lightweight, and useful.

Sandy's favorite

Model: SWISSMEMORY® USB Storage in a VICTORINOX Pocket Knife 128MB
Manufacturer: Victorinox
URL: www.victorinox.com
Price: $61.00

Executive Attaché

This is one of the most useful devices ever. If you need to carry your files between work or school and home, you want to show your pictures to a neighbor, or you want to make a quick backup of your files, just take out your Executive Attaché. It's a pen and a USB hard drive. Just unscrew the top of the pen to reveal the USB drive. Plug it into your computer, and you have a portable drive. The Attaché works with both PCs and Macs. You always need a pen, and computer files are becoming an indispensable take-along.

Model: Executive Attaché, 256MB
Manufacturer: PNY
URL: www.pny.com
Price: $29.99

SanDisk Cruzer Profile

SanDisk's Cruzer is a tried-and-true USB portable hard drive. I've been using mine for more than a year. Just plug it into the USB port, and the device shows up as a drive in the My Computer window. Copy files to the USB drive, and you have an extra backup copy of your files to carry with you wherever you go. The Cruzer works with both PCs and Macs.

SanDisk has added one really neat feature to the Cruzer with the Cruzer Profile, which includes

Model: 512MB Cruzer Profile
Manufacturer: SanDisk
URL: www.sandisk.com
Price: $99

fingerprint recognition technology. The device, which is about the size of a pack of gum, has an integrated fingerprint reader. After you have set up the Cruzer software and added your fingerprint, you only need to swipe your finger across the sensor to activate the drive. The Cruzer Profile also comes with a data file encryption program, which synchronizes with Microsoft Outlook, and a program that allows for file backups with minimal memory consumption. A larger 1GB unit is also available.

Iomega REV Backup Drive

One of the first rules of computing is to always back up your files. You can back up on CD, but CDs are not large enough to back up your entire computer. This Iomega Rev drive provides a way to back up your whole hard disk.

This starter kit comes with the Rev unit plus six high-capacity disks. The unit attaches to the PC with a USB cable. Excellent backup software is included. Each disk holds up to 90GB, so you can back up your entire computer on one disk. With six disks, you can use a different disk each time you back up and always have an extra backup for those times when everything seems to go wrong.

Model: REV 35GB/90GB Starter Kit
Manufacturer: Iomega
URL: www.iomega.com
Price: $579.99

You can use the small rigid disks over and over. The included Iomega Automatic Backup Pro software allows you to back up your entire hard drive or just your important files. It is easy to set up and easy to use, giving you secure and safe data backup.

Freecom USB

Most USB hard drives are small, chunky devices that can be attached to a keychain or worn around the neck. The Freecom card is different. It looks like a credit card and can easily be stored in your wallet with your other credit and debit cards. The USB connector folds out of the card. Insert the connector into the USB port, and you have a USB hard drive to store computer files, audio files, and pictures. The connector rotates to accommodate the different types of computers and locations of USB ports. The Freecom card has a much higher capacity than most USB hard drives. It comes in sizes up to 60GB and includes software that takes a complete image of your hard drive for backup purposes. The Freecom works with both PC and Mac computers. With fast transfer rates, if you need to move or store a lot of data or want a quick way to back up your hard drive, the Freecom card is an excellent choice.

Model: FHD-XS 20GB Freecom USB Card
Manufacturer: Freecom Technologies
URL: **www.freecom.com**
Price: $159

Griffin PowerMate

One of the coolest-looking little gadgets is the Griffin PowerMate. It is a USB device for PC or Mac that is basically one big knob about 2 inches in diameter. It comes in matte black or brushed aluminum finish. With the included software, it can perform a variety of functions. It can be a knob to control the volume of your computer's audio—handy when the

Model: PowerMate
Manufacturer: Griffin Technology
URL: **www.griffintechnology.com**
Price: $45

boss walks in or the telephone rings. You can also use it to scroll through web pages. It works great to cycle through projects when you are creating or editing audio or video. You can even program several unique settings and have the PowerMate remember them all. Although the functionality is certainly worthwhile, it is the looks of this gadget that will turn heads. Made of heavy machined aluminum, it is strong and rich looking. When you turn it, a blue light emanates from the bottom, growing stronger as you turn the dial. You might want to buy it just to look at it!

Zip-Linq Retractable Cables

If you hate the gaggle of cords leading to your computer, you'll love Zip-Linq. These retractable cables roll up inside a small, round cable protector. A 48-inch cable retracts to less than 4 inches long. After my grandson played with one for several hours, I declared the Zip-Linq to be sturdy *and* childproof. The USB cable can connect any USB device to your computer. The Zip-Linq cables also come in USB to Mini plugs that are needed for digital cameras, telephone car chargers, FireWire cables, and a variety of other cables. There is even a portable Zip-Linq mouse with a retractable cable, which is perfect for travel.

Model: USB 2 A-B Cable
Manufacturer: Zip-Linq
URL: www.ziplinq.com
Price: $7.99

Linksys Wireless Router with Speed Booster

When you're hooked on computers, you almost always have to have more than one. The easiest way to share an Internet connection, files, and printers between those computers is to create a wireless network. With a wireless network, you don't have to run cables through or around walls.

Courtesy of Linksys

The main component of a wireless network is the router/access point. This connects to the broadband modem that usually comes with your broadband cable, satellite, or DSL (Digital Subscriber Line) connection. This Linksys router connects to the newer 802.11g

Model: WRT54GS
Manufacturer: Linksys
URL: **www.linksys.com**
Price: $99

wireless networks and the older 802.11b networks. Its Speed Booster technology increases wireless performance up to 35%. It features 128-bit encryption and has a built-in firewall for added security.

RangeMax Wireless USB

The RangeMax is a simple USB adapter. Just plug it into any USB port, and the Smart Wizard install assistant walks you through getting the

attached computer on a wireless network. This 802.11g device works with all wireless routers and access points and also works with the older 802.11b networks. The best thing about this adapter is that you can use it for both desktop and laptop computers. You don't have to open the desktop computer to install a card. Just plug this in instead.

Model: RangeMax Wireless USB 2.0 Adapter WPN111
Manufacturer: NETGEAR
URL: **www.netgear.com**
Price: $72

Fingerprint Reader

It might sound like a dream, but it's here now. Whenever you visit a website where you usually have to enter a username and password, you simply put your finger on the fingerprint reader, and this device automatically recognizes you and enters your username and password. You can also use it to log on to Windows. For even more security, you can set it so that you, and only you, can start up your computer. This product is for Windows XP, Windows Tablet Edition, or Windows Media Center Edition only.

Model: Fingerprint Reader
Manufacturer: Microsoft
URL: www.microsoft.com
Price: $54.99

OTT-LITE VisionSaver Plus® DeskPro Task Lamp

It is difficult to get the proper lighting when working on the computer. Lamps and overhead lighting can cause a glare on the screen. If the computer screen is properly lit, often there is not enough light to read the papers next to the computer. The OTT-LITE VisionSaver Plus® DeskPro Task Lamp is perfect to use when working on the computer. Small and compact, the fold-up design and height-adjustable head makes it easy to move the light to get it in the proper place. The lamp itself can be adjusted so that it illuminates papers and books on the table or desktop but does not cause a glare on the computer screen. Best of all, the OTT-LITE VisionSaver Plus light mimics natural daylight, making it easier to see and easier on your eyes.

Sandy's favorite

Model: OTT-LITE VisionSaver Plus® DeskPro Task Lamp J59376
Manufacturer: OTT-LITE Technology
URL: www.ottlite.com
Price: Valued at $79.95; retails for $39.95

Protective Equipment

Lightning and power spikes can damage electronic equipment. If you think that lightning won't come near your home, think again. Lightning strikes somewhere on the surface of the earth about 100 times every second. The United States has more than 100,000 thunderstorms a year. Even when no storms are in sight, power surges in the electric lines are a common occurrence.

The way to protect your computer from lightning and power surges is to use a good surge protector. Don't be confused. A surge protector is not simply a strip of outlets. A good surge protector is made up of a series of metal-oxide varistors (MOVs) that shield the computer from abnormally high voltage. When your home receives voltage spikes of high intensity, the MOVs grab the current and push it away from the computer.

A good surge protector should offer four features:

- It should protect against lightning strikes. Some do not.

- It should offer insurance to cover the loss of properly attached equipment.

- If you have a regular modem, your surge protector should have an R-11 telephone jack where you can hook up your telephone line.

- If you are using a cable or satellite Internet connection, your surge protector should also accommodate your television/Internet cable.

Performance SurgeArrest

This surge protector has it all. It has 11 outlets, six of them widely spaced to accommodate power blocks. Telephone, cable, and wired network connections are all included. The unit comes with $100,000 in insurance coverage for connected equipment. The SurgeArrest has connectors that allow you to hook up two telephone lines, an Ethernet network cable, and

Model: Performance SurgeArrest 11 Outlet with Tel2/Splitter, Coax, and Ethernet Protection
Manufacturer: APC
URL: www.apc.com
Price: $39.99

a television/Internet coaxial cable. The unit guarantees protection from lightning and power surges.

It also includes some nice extra features such as a 180-degree swivel on the 10-foot power cable, a plastic guide to help manage cords, three always-on outlets, and safety shutters that protect users from accidental contact with unused outlets. There is even a yellow light that goes on if your connected equipment exceeds the unit's capacity. If you don't have a wired network or a cable connection, you can choose a different model that is less expensive. Just make sure that what you choose has protection against lightning and insurance to cover your equipment.

Tripp Lite Home Theater UPS

A surge protector protects your equipment from power surges and nearby lighting strikes. This Home Theater unit gives you that protection plus much more. It is a UPS, so it provides battery backup power that allows you to continue working and to back up your files in case of a power shortage or blackout. This Tripp Lite product has some features that make it exceptional. First, it covers all connected equipment with lifetime insurance of up to $100,000. Other great features include up to three hours of battery backup power. The eight power outlets are well-spaced, allowing you to plug in equipment with large transformers. The unit also has inputs for a telephone line to cover dial-up Internet connections and pay-per-view satellite connections that use a phone line. In addition, your Internet or television cable line can be connected to this device to protect your

Model: HT 1500 UPS
Manufacturer: Tripp Lite
URL: www.tripplite.com
Price: $169.99

equipment from power surges that might come down the cable line. The unit also suppresses line noise, resulting in sharper pictures and clearer audio.

This UPS unit is called the Home Theater UPS because it can easily protect your cable, satellite, television, DVD player, and literally all of your home theater components, besides being great for your computer. The software even provides for automatic unattended shutdown of your computer. It's perfect for a media center PC, a computer setup, or your home theater equipment.

SurgeArrest Notebook Surge Protector

If you use a laptop for business or travel, don't forget to protect your computer when you take it on the road. Several types of surge protectors are available for laptops. This SurgeArrest is called in-line because it plugs directly into the power block on your computer. You can choose between a two-prong or a three-prong unit based on the design of your laptop computer. This compact device is perfect for travel. It protects the telephone and network line in addition to the power line. It is guaranteed against lightning, surges, and spikes and comes with $75,000 insurance for properly connected equipment.

Model: Model PNoteProC8
Manufacturer: APC
URL: www.apc.com
Price: $19.99

Email and Internet Hardware

Email is one of the most popular online activities. It brings family and friends together from all over the world without the struggle of writing letters or incurring large telephone bills. Most people access email through computers. Some folks, however, don't want the hassle of a computer. For them, several email devices are available that are easy to set up and use.

PocketMail

PocketMail Composer is a simple email device. It's not for everyone. If you have a laptop computer or an email-enabled PDA, PocketMail seems like a step backward. It is, however, an uncomplicated way to have access to email if you don't have a more sophisticated mechanism. The device and the service are well-designed, easy to learn, and easy to use.

Toll-free numbers are available for the United States, Canada, and several foreign countries. The PocketMail folks even printed the toll-free numbers on the back of the unit, which is useful. Although you cannot receive large email attachments on PocketMail, you can view attachments at the PocketMail website for 30 days.

Model: PocketMail Composer
Manufacturer: PocketMail
URL: http://www.pocketmail.com
Price: $99 plus monthly service

PocketMail Composer is powered by two AA batteries. It needs no other power and has no cords or wires, so it is mobile. I know several people who use it when they take to the road in their RVs. Besides being an email device, you can use the integrated keyboard to turn the Composer into a Personal Digital Assistant (PDA) with an address book, calendar, scheduler, memo pad, and calculator. You hold the PocketMail device up to the telephone to dial. The technology is a little dated, but it is a good way to get started if you have never used email before. Monthly service is $149 for 12 months.

MailBug

Like PocketMail Composer, MailBug allows you to communicate by email without a computer. It is a free-standing device with a standard size keyboard and an adequate size screen that hooks up to your regular telephone line. MailBug is easy to set up and easy to use. It has arrow and scroll keys for navigation and function keys for accessing features that are listed on the screen. One nice feature is that you can send faxes from your MailBug. Another interesting

Model: MailBug
Manufacturer: Landel
URL: **www.mailbug.com**
Price: **$125 plus $9.95 per month service**

aspect is that you can access your MailBug email from any Internet-enabled computer. The Bug is easy to use and comes with full instructions. It also provides access to news, stock quotes, sports scores, weather, and other online information, but it can't handle email attachments. MailBug service is $9.95 a month.

MSN TV 2 Internet & Media Player

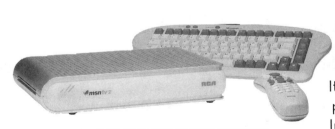

The MSN TV 2 Internet & Media Player is the latest rendition of the MSN TV, which was previously called WebTV. It is the closest thing to a computer when it comes to Internet access. You hook the MSN TV unit to your television and your telephone line. Then you use the wireless remote control and wireless keyboard to access the Internet and send and receive email. You can surf

Model: MSN TV 2 Internet & Media Player
Manufacturer: Microsoft
URL: **www.msntv.com**
Price: **$199 plus monthly service**

and shop just like you do when using a computer. You can view photos, enjoy music, play video clips, and even use instant messaging. The only drawback to MSN TV is that you must navigate without a mouse, which is inconvenient to anyone who has already seen the ease of use that a mouse provides.

The MSN TV uses dial-up telephone access. If, however, you already have a broadband home network, you can purchase the MSN TV 2 Internet & Media Player and use it as a secondary Internet access point. With this setup, you can show videos and pictures from any computer in your home on your television. You can also listen to music stored on any PC on your television. It is somewhat like having another computer for Internet, music, slideshows, and videos. Even though you might already be paying another Internet service provider (ISP) for your broadband access, to use the MSN TV 2, you must pay an additional $9.95 a month for the MSN service. Dial-up service is $21.95 a month.

MailStation

The wireless MailStation is an email-only device that comes with EarthLink dial-up service. The unit is easy to set up and use. At 11" × 9" × 1.5", the MailStation doesn't take up much space, and it's good looking enough to fit in any room of the house. At less than two pounds, it's lightweight and portable. The MailStation has a jog dial and a Select button that make choosing functions easier than using a mouse. It also has five large buttons under the screen for making selections. The MailStation even gives you one-button access to weather for your area and news from around the world. Drawbacks are that only old DOS printers can be attached. MailStations are still being sold and supported by EarthLink. EarthLink has, however, stated that it will not be actively selling this product in the future. Even so, MailStations are viable email devices. They are currently available at Best Buy and some other retail outlets.

Model: MailStation 350
Manufacturer: EarthLink
URL: www.earthlink.com
Price: $199, plus monthly service

Voice Email

If you love email but hate to type, this little device lets you record your voice and email the recording without typing. The small Voice Email fits nicely into your hand. The built-in microphone has one-touch recording with voice activation. You simply record your message and then hook up the recorder to your computer with the included USB cable. The included software transfers your voice to the computer and converts it into a file. The software even starts up your email program and attaches the file so that all you have to do is insert the recipient's name.

The Voice Email can record up to seven hours, so you can also use it to record lectures or to make notes to yourself. It's small and portable, so you can record notes

Model: Voice Email DV8
Manufacturer: Datexx
URL: www.datexx.com
Price: $110

and email anywhere. Unfortunately for Mac users, this device only works with Windows XP and Windows 2000.

Sandy's Lingo List

The world of technology has created some crazy new words. Here are explanations for a few of the more unusual words used in this chapter.

broadband—A high-speed Internet connection that is always on, so there is no need for the modem to dial to establish an Internet connection. For most average users, a broadband connection accesses the Internet by a cable modem provided by their local cable company, a DSL modem and telephone line provided by their local telephone service provider, or a satellite connection.

FireWire—A way to connect external devices that is also known as IEEE 1394. FireWire is faster than USB and supports data transfer rates of up to 400Mbps. Because of its high speed, FireWire is often used for transferring video files. FireWire was originally developed by Apple. It is found on most Apple computers and some PCs.

router—A piece of equipment that moves data in a network.

uninterruptible power supply (UPS)—A power supply that includes a battery to maintain power in case of a power outage or interruption. A UPS, powered by its own battery, keeps equipment plugged into it running for a certain period of time after the regular power has been interrupted. Typically, a UPS also functions as a surge protector.

universal serial bus (USB)—A way of connecting external devices to a computer. USB allows data to be transferred between the connected device and the computer much faster than older serial or parallel connections. A USB device is easy to install. On most newer computers, you just plug the device into the USB port on the computer, and it works without additional software. All Macs and PCs sold in the past several years have USB ports.

wireless—A generic term that refers to any nonwired transmission of data. Although this includes radio, television, and cell phones, it is most often used to refer to nonwired networked computers and devices, such as keyboards and mice that are not attached to the computer by wires.

Chapter 3

Entertainment Gadgets

Edison did not invent the first talking machine—he invented the first one that could be turned off.

—Anonymous

Many of us grew up with radio. Some of us grew up with vinyl records and black-and-white television. We have all watched technology bring us new and better ways to entertain ourselves. We now have satellite radio, digital video recorders, portable music players, and noise-canceling headphones. The array of gadgets and gizmos that were made purely for our listening enjoyment is astonishing.

Audio Gadgets

We have been enjoying radio, which was our first audio entertainment device, since the early 1900s. Radio is still a major entertainment device, but even it has taken on new dimensions. We can now choose from more than 100 stations of radio programming that is beamed off of a satellite.

Many of the other audio devices available today weren't even dreamed of just a few years ago. I imagine that Thomas Edison or Henry Ford would have been amazed at a handheld music player that contained more than 1,500 songs. In fact, you might also be amazed by some of these great high-tech audio devices.

Delphi MyFi xm2go Satellite Radio

With satellite radio, you can travel from New York City to the California coast listening to the same commercial-free radio station during the whole trip. XM

Sandy's favorite

radio has more than 150 channels of radio programming. You can choose from talk, entertainment, traffic, weather, and just about every genre of music that you can imagine. There is music from the 40s, 50s, and 60s, in addition to country, jazz, rock, blues, Latin, Christian, and more. XM currently has NASCAR and major league baseball coverage, too.

MyFi brings you the best of satellite radio. It is a handheld unit about the size of a PDA that comes with a home adapter kit, a vehicle adapter kit, antennas, carrying case, remote control, headphones, rechargeable battery, and car mounting kit. You can move it between work, home, car, or boat. At home or the

Model: MyFi xm2go
Manufacturer: Delphi
URL: www.xmradio.com
Price: $299; XM service $12.95 per month

office, the device hooks up to a radio or stereo system. On the road, it uses a cigarette lighter and the car or truck radio. Because the MyFi allows you to record satellite programming, you can also use it as you would a portable music player to listen to any shows or music you have recorded.

In addition to the price of the MyFi unit, you need to pay a monthly subscription fee. Keep in mind that XM has a lot of programming for the price.

Starmate Sirius Satellite Radio

The Sirius satellite radio service is similar to XM only with different programming. Like XM, you can use the Sirius service to listen to the same commercial-free radio station anywhere in the country. Sirius has more than 120 channels with music of every genre, sports, news, original programming, and top news and talk radio personalities. The sports coverage includes the National Football League, NBA basketball, and NHL hockey.

The Starmate unit is compact and portable and comes with everything you need to hook it up to your car radio. You can also use it in your home or office with the optional home kit. The three-line display is a readable red color. The well-designed unit features easy-to-use buttons and comes with a remote control.

Model: ST1
Manufacturer: Sirius
URL: www.sirius.com
Price: $99.99 plus service, $12.95 per month

MuVo Micro

At 1.32" × 2.58" × 0.51" and 2.2 oz. with the battery, the MuVo Micro, also known as the Zen Nano Plus, is so small that you barely know it's there. This rectangular music player comes in eight different colors including vibrant colors like pink, orange, and yellow.

The backlit LCD display is small but clear. It allows you to see what song is playing and to easily find the song or artist that you want to hear. The MuVo is perfect for running and working out in the gym. You can keep it securely on your arm with the included armband, or you can clip it on your belt. The integrated FM radio allows you to listen to the gym's TV when running on the treadmill or catch your favorite radio show. The built-in microphone allows the unit to be used as a voice recorder. The MuVo comes in several memory capacities up to 1GB.

Model: 256MB
Manufacturer: Creative Labs
URL: **www.creative.com**
Price: $79.99

If you love audio books, you can get the 128MB version of the MuVo player for free with a year's subscription to the Audible.com service. The Audible.com website has more than 25,000 books, radio shows, newspapers, and magazines to choose from. You can subscribe to one book per month for $14.95 or two books per month for $21.95. You simply download the books to your computer and transfer them to your MuVo or other audio player or burn them to a CD. The process is easy, and the selection is excellent. If you are a book lover, this is an inexpensive way to listen to audio books.

Apple iPod

The Apple iPod has become a wildly popular portable music player because it offers a simple, easy-to-use interface and storage for a lot of songs. This iPod's 60GB hard drive holds up to 15,000 songs. The clear screen and scroll wheel make it easy to choose the music you want to play. The iPod can also be used to store photos. Up to 50,000 pictures can fit on this small device. You can view the photos on the 2-inch color LCD screen. The iPod has now joined the podcasting revolution. With the latest iTunes software, the iPod has built-in support for Internet radio broadcasts called *podcasts*. You can sync your iPod and listen to podcasts from your favorite websites.

The iPod is small (.75" × 2.4" × 4.1") and lightweight (6.4 oz) and has outstanding battery life. The iPod comes with earbud headphones, AC adapter, USB cable, and iTunes software. Although this is an Apple product, it works with both PCs and Macs.

Model: iPod 60GB
Manufacturer: Apple
URL: www.apple.com
Price: $399

Archos Gmini 400

The Gmini is a digital audioplayer that doubles as a photo display device and a portable video player. It has a 20GB hard drive and a 2.2" color LCD display. You can use the included cables to connect the Gmini to your television to play video or to your stereo system to play music. You can also use it to play games or to copy photos directly from a compact flash card. Connect the Gmini to your computer, and you can copy pictures, audio, or video to the device. It is excellent as an audio and video player, but it's only average for photo display.

As you might expect from a do-it-all gadget like this, you need to read the manual and search through the various menus to learn all the functionality of this device. There are myriad menu choices and an almost dizzying array

Model: 400
Manufacturer: Archos
URL: www.archos.com
Price: $399

of features. The Gmini has a lot of software preinstalled, but you'd never know it was there unless you searched through the files and folders on the device.

Although this is a cool device, it will take a time investment to learn all of its functionality.

iriver H10

iriver has come up with a good alternative to the iPod portable music player. At almost 5 inches tall, the H10 is a little larger than many other players, but it feels good in your hand, and at only .5 inches thick, it can easily slip into your pocket. Like the iPod, it has a minimum of buttons and a good clean interface that is easy to navigate. The color 1.5-inch LCD is clear. It has a 6GB hard drive that holds a ton of photos or songs and has good audio quality.

The H10 has several nice features that other music players lack. Many portable music players like the iPod have a rechargeable battery that can only be replaced by sending it back to the factory. This iriver has a swappable battery, so you can buy a spare or easily replace it if it wears out. The H10 comes in

Model: H10
Manufacturer: iriver
URL: www.iriver.com
Price: $279.99

red, blue, silver, and gray. It comes with a wonderful soft molded plastic case with belt clip, and it can double as a voice recorder. A built-in FM radio lets you listen to and record radio, too.

SoundBridge

The Roku SoundBridge connects to your stereo system to play any music that you have stored on your computer. It can use a wired Ethernet connection or a wireless connection to access the music from your computer. SoundBridge uses Windows Media Connect or iTunes to serve up the music. The SoundBridge is compatible with Windows

Media Player 10 and many online music stores such as MSN Music, Napster, iTunes, MusicMatch, and CinemaNow.

Model: M2000
Manufacturer: Roku
URL: www.rokulabs.com
Price: $399

The elongated aluminum body of the SoundBridge is good looking. The bright blue 12-inch wide display is clear and visible from any angle. You can even see it from across the room. You can use the remote control to browse music by song, album, artist, composer, or genre. You can also use playlists that you have created in iTunes or other music software programs.

Wurlitzer Digital Jukebox

Who doesn't remember the Wurlitzer jukeboxes of the 1950s and 1960s? We slid a few coins into the jukebox and swayed while we ate hamburgers or danced the night away. Now the Wurlitzer is available to play your CDs and digital music.

There are no bubble lights around the jukebox, but the design is simple and elegant. It includes an integrated subwoofer and two tall and lanky speakers from Klipsch. The digital jukebox can hold up to 1,000 CDs. The beauty of it is that all you do is insert the CD you want to copy into the jukebox. The music, track information, titles, genre, and digital cover art are automatically stored in the jukebox. After you have inserted all your CDs, you can pack them up and store them because you won't need them anymore to play your music.

Model: Freestanding
Manufacturer: Gibson Audio
URL: www.wurlitzerdigitaljukebox.com
Price: $1,999

You access the music via the touchscreen remote control. The interface is simple and easy to use. This jukebox also has a variety of options. You can export songs to a portable music player. Or you can add other Wurlitzer receivers to distribute music wirelessly throughout your home. You can access the Wurlitzer Music Service through a wireless connection to your home network. With this, you can access many channels of digital radio and a download store where you can purchase thousands of songs and send them directly to the jukebox. A smaller version without the speakers is also available.

Sonos Digital Music System

If you have a lot of digital music stored on your computer, the Sonos digital music system lets you play that music anywhere in your home and control it from a small handheld remote. The Sonos system does it all, so you don't need a PC in every room, a music server, or a wireless network. The Sonos zone player that comes with the system distributes, plays, and amplifies the music. The Sonos controller is a wireless remote control with a full-color bright LCD screen, a scroll wheel control, and simple buttons.

You install the first zone player by attaching it to your PC via the Ethernet connection. Then you can hook up another zone player to a set of speakers in any room in the house. You can have up to 32 zone players. The installation is easy, and after it's installed, you can play any music files on your PC, Internet radio, or music from external audio sources like CD players and portable MP3 players.

Model: Sonos Introductory Bundle
Manufacturer: Sonos
URL: www.sonos.com
Price: Controller and 2 Zone Players, $1,199

The controller manages the audio of all the zone players. You can play a different song in each zone or the same music in all zones. You can also control the volume for each zone separately right from the controller. Sonos isn't cheap, but it's a great concept, and it works well.

Griffin Radio SHARK

What TiVo is to television, Radio SHARK is to radio. All you need to do is install the included software and plug the Radio SHARK into the USB port on your computer, and you are ready to trawl the radio waves. You set the radio station via the onscreen menu and listen to any program on your computer speakers. The software also allows you to easily record any program or series of programs.

The "time-shift" recording is cool. When you are listening to a radio show, you can press Pause to get a sandwich or answer the door. When you return, you simply press Play again, and the programming resumes right where you left off.

Model: Griffin Radio SHARK
Manufacturer: Griffin
URL: **www.griffintechnology.com**
Price: $69.99

Just in case you're wondering, the SHARK gets it name because it looks like a large white fin. The lighted blue wave design on the fin turns cherry red when recording. The SHARK works with both PCs and Macs and provides hours of pleasant radio listening.

QuietComfort 2 Acoustic Noise Cancelling Headphones

These lightweight headphones fold for storage and come in a convenient travel case. The headphones cover your entire ear with a padded cup. They are adjustable and quite comfortable.

The beauty of these headphones lies in the sound quality and noise cancellation

Model: QuietComfort 2 Acoustic
Noise Cancelling Headphones
Manufacturer: Bose
URL: www.bose.com
Price: $299

effects. You can wear them for air travel to block out excess noise and, of course, you can attach them to your favorite music player to enjoy personalized music at home or on the go. The headphones come with adapters and extension cable so that you can easily hook them up to your home stereo system.

Sennheiser Earbuds

The earbuds that come with many portable music players are pretty poor quality. Adding a good pair of earbuds can improve the sound quality of almost any player. Sennheiser is known for its sound quality, and these little earbuds don't disappoint. The buds are a cool metallic blue color. I hate wires hanging every-where, so I really appreciate the convenient windup case that holds the excess cable. I also like the volume control that is built right into the cable for easy access.

Model: MX500
Manufacturer: Sennheiser Electronic
Corporation
URL: www.sennheiserusa.com
Price: $19.95

Digital Vinyl CD-R

Bringing the old and the new together, these writable CDs look just like old 45-rpm records, complete with vinyl-like grooves. The retro look is

just plain fun when you want to create and share music CDs. You can even use them when you decide to digitize those old LP records. What fun!

If you own a printer that can print on CDs, there is a special coated label area that you can use with an inkjet printer to create really impressive-looking labeled CDs.

Model: Digital Vinyl CD-R
Manufacturer: Verbatim
URL: www.verbatim.com
Price: 50-pack $49.99

Verbatim's special dual protective layer technology ensures that your CDs have enhanced protection against high temperatures, humidly, sunlight, and rough handling. Verbatim states that these CD-R disks have an estimated archival life of more than 100 years.

Video Equipment

Video is everywhere today. You can find it on the Internet. Many portable devices now display video and audio. People are watching videos on the train, in the car, and even in the doctor's office. Great video devices add a lot of entertainment value to our lives.

Archos AV420

The Archos AV420 is a video-music-photo player that can also record from just about any source. It comes with a cradle that you can attach to your TV, DVD player, VCR, or CD player. Then you simply drop the AV420 into the cradle when you want to record.

Sandy's favorite

There is a lot to like with this media player. It has a great screen, excellent video replay, a wireless remote, a removable battery with excellent battery life, and it supports TV recording with a program guide. It comes with a carrying case, but it is small enough to fit in a purse or baggy pants pocket. Also, the Archos has a built-in speaker so that you don't have to use the included ear buds unless you want to.

Model: AV420
Manufacturer: Archos
URL: **www.archos.com**
Price: $550

The Archos is an excellent jack-of-all-trades. It offers a built-in microphone for live audio recordings and can double as a portable hard drive. The CompactFlash memory-card slot allows you to use it to store photos from a digital camera when you are on the road. Copying videos from your computer to the Archos is a bit complicated, yet this player is a winner. And it works with any version of Windows and with the Macintosh.

ForceField DVD

DVDs were made to be handled with care. To prevent fingerprints, you should only touch the edges. To prevent scratches, you should handle them with care. To keep them dust free, you should keep them in a case. The problem is people don't always adhere to these care instructions. They throw disks in piles, drag them across desks, and leave them lying around to accumulate dust.

Model: ForceField DVD 30-Pack
Manufacturer: Imation
URL: **www.imation.com**
Price: $39.99

Imation has found an answer to this type of treatment by creating disks with a special ForceField scratch-resistant coating. Standard DVDs are made with a polycarbonate cover layer on the recording site of the disk. The ForceField DVDs have an extra polymer layer of protection on the recording side of the disk. This provides more resistance to scratches, smears, and dust. Imation ForceField CDs are also available.

Bose 3-2-1 Home Entertainment System

The 3-2-1 provides theater sound from a compact home stereo system. The package includes two Gemstone speakers that are small enough to fit in the palm of your hand. The larger Acoustimass module can be hidden behind a piece of furniture. Because there are no rear speakers, you don't have to run wires to the rear of the room.

Installation is easy, and sound quality is exceptional. I was blown away when I watched my first movie with the surround sound system. The Bose 3-2-1 includes a CD/DVD player, AM/FM runner, and universal remote control that you can program to run most of

Model: GS Series II DVD
Manufacturer: Bose
URL: **www.bose.com**
Price: $1,299.00

your entertainment devices in addition to the 3-2-1 system. The system is not cheap, but the quality is excellent, and technical support is great.

ZVUE

The ZVUE is a portable music and video player plus a photo viewer. At 4.33" × 2.91" × 1.1", it feels a little chunky in the hand, but the audio and video capabilities are good for the price. The 2.5-inch TFT color screen is bright and clear. It comes with two stereo headphone jacks so that two people can use it at the same time. Controls and menus are easy to navigate.

Although most players of this type use a rechargeable battery, the ZVUE uses four AAA batteries, which last for about eight hours of video playback. Batteries can get expensive, but this can also be a useful feature. When the batteries run out, you simply insert new ones without having to wait for recharging.

The ZVUE comes with a USB cable and the software necessary to put your own videos on the player. Moving videos and music to the ZVUE is easy. It also comes with a 32MB memory card, or you can purchase a ZVUE with a 512MB card for $159.95.

Model: ZV003
Manufacturer: Handheld Entertainment
URL: www.ZVUE.com
Price: $99.95

DigitalMovie DVD

Another link to the past... The Verbatim DigitalMovie DVDs look just like old-fashioned movie reels. They are available in several speeds and formats. The Verbatim DVDs use a high-performance metal azo recording dye and boast superior archival stability, but it is the look of these DVDs that is so enticing. They exude nostalgic charm.

Use them for giving out videos of that special wedding video or to transfer your VHS tapes to digital format.

Model: 4.7GB 4X DVD +R
Manufacturer: Verbatim
URL: www.verbatim.com
Price: 3-pack $9.99

Television Devices

I will always remember the first time that I saw television in color. The image of a ballet dancer in a pink tutu stays with me even today. Television is still one of our basic entertainment devices, but now we use many television-related gadgets and gizmos to enhance our television viewing. Today we can record every instance of our favorite television show to watch any time we please. We can even watch and record television on our computers. In fact, if you read on, you'll find that we can even watch our home television from the other side of the world. These exciting television devices augment our viewing pleasure.

TiVo Digital Video Recorder

TiVo is like a digital VCR. The move to digital has many advantages. Recordings are saved on the unit's hard drive, so there are no tapes to fumble around with. It's also much easier to use TiVo than those old VCRs.

Model: Series2; 140-hour recording
Manufacturer: TiVo
URL: **www.tivo.com**
Price: **$299 plus $12.95 per month for service**

TiVo comes with its own easy-to-use onscreen programming guide. You tell it to record every instance of your favorite television show, and it automatically finds every instance and records it. You can do the same with actors, actresses, and directors. You can also record every game of your favorite team or every NASCAR race. Another great feature is the ability to pause live TV. If the doorbell rings or you need to get a drink, press the Pause button; when you return, you can see the show from right where you left off. Also, TiVo gives you the ability to replay TV. If your favorite player makes a touchdown, you can replay it with the touch of a button.

TiVo even lets you add the TiVo device to your home network. With this capability, you can schedule your TiVo recordings online. So if you are at a friend's home and realize that you forgot to record *Desperate Housewives*, you can go online and schedule the recording. You can use the TiVo combined with a home network to burn DVDs, play your digital music, and view photos.

You can use TiVo with a regular television with antenna, cable, digital cable, or satellite television. The most expensive version lets you record 140 hours of video. Cheaper versions have smaller hard drives and can hold less video. Setting up the TiVo takes some time, but the results are worthwhile. The only downside is the additional monthly fee.

Terk Volume Regulator

This is one of the most useful gadgets I've found. I am constantly aggravated by television commercials that are louder than the programming and some television stations that are louder than others. Instead of constantly adjusting the volume, you can simply have this device regulate the volume for you. It is also great for the hearing impaired because it raises the sound level of hard-to-hear programming.

The volume regulator uses standard audio cables to connect to the television. After it's connected, it works automatically with no user intervention. It samples the sound quality thousands of times per second and uses advanced digital signal processing to automatically adjust the sound levels. The consistent sound levels that it produces are immediately apparent and welcome.

Model: VR-100
Manufacturer: Terk
URL: **www.terk.com**
Price: $49.99

Universal Remote

If you want to coordinate your entire home theater, this Universal Remote is just what you need. You can customize all the functions, including the labeling of functions that appear on the LCD panel.

The Universal Remote can take the place of all your other remote controls. Setup is fairly easy with a detailed tutorial on the included DVD and a well-written user manual. This remote works by both infrared (IR) and radio (RF) signals to give a great range that allows it to work even when physical obstructions are a problem. With the addition of a $75 extender, the remote can work throughout the house and even outdoors up to 100 feet.

I like the touch screen, which is clear and well designed. The buttons could be a bit larger, but they are well placed and easy to use. The buttons and screen are backlit to help with evening visibility. Although it is a bit more complicated, you can also program the remote to control multiple devices with a single button. If you need to switch the input on the television when you play the DVD, this remote lets you perform both tasks with one button.

Model: URD-300 Customizer
Manufacturer: Universal Remote
URL: www.universalremote.com
Price: $200

TV Remote with Alarm Clock

Do you hate to miss your favorite television show? This gadget makes sure you remember. It has three TV alarms and one buzzer alarm. You follow a few simple steps to "train" the device to work with your television remote. Then set the television alarm(s), and your TV starts at the right time so that you don't miss your favorite shows.

The small unit is good-looking and stands on its own, so you can use it as a small clock, too. It works on AA batteries and has a button that backlights the display for night-time viewing.

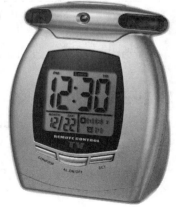

Model: 3010
Manufacturer: Digitt
URL: www.alldmd.com
Price: $9

Media Center PC

Sandy's favorite

Before you buy your next computer, you might want to check out a Media Center PC. These PCs are made by many different manufacturers and run a special version of Microsoft Windows called Windows XP Media Center Edition. This version of Windows is available only on a new PC.

The Gateway unit is a fully functional and fully capable PC with a Pentium D 2.8GHz processor, a 250GB hard drive, 1024MB memory, writable DVD, 8-in-1 media card reader, keyboard, mouse, and speakers.

Although you can use it for all your computing needs, this media center PC really shines as a music, photo, and video center for the entire family. This computer has special software that helps you download, copy, and organize your digital music. With the built-in card reader, you can take the media card from your digital camera and insert it into the computer for an instant slide show. The computer also has a built-in television card that allows you to watch and record television shows, much like a TiVo. Although

Model: 835GM
Manufacturer: Gateway
URL: www.gateway.com
Price: $849.99

TiVo charges a monthly fee for its service, television recording on a Media Center PC has no monthly fee. You can easily program it to record one program, a series of programs, or every instance of your favorite show. You can also hook it up to a television to display the television programming you have recorded.

The Gateway 835GM comes with a keyboard and mouse for use with the computer, but it also has a remote control that can be used for the Media Center functionality when showing pictures, playing music, and recording or playing videos. It's a great idea, and it performs brilliantly.

Big Button Universal Remote Control

The Big Button Remote Control is three times larger than most ordinary remote controls. Each button is clearly marked and lighted. It is shaped like a T-bone to make it easier to hold. It can control up to five pieces of equipment, including the TV, VCR, DVD, cable, and satellite.

Another advantage is that this remote is so big that you won't easily lose it.

Model: Big Button Universal Remote Control
Manufacturer: Gold Violin
URL: www.goldviolin.com
Price: $39

DVD/VCR Combination Unit

Did you know that VHS tapes are rated to last only 15 years? If you want to preserve all those memories that you have on tape, the best thing to do is to transfer them to DVD. You can do that many ways, but the absolute easiest way is to purchase a combination VCR and DVD device. With the Sony VX500, you simply insert the VHS tape and a blank DVD, and the entire process is done with the press of a button.

You might know that several different DVD formats are available, with the plus (+) format and the minus (–) format both still viable options. Because the VX500 records in both formats, you can actually

transfer the information from your tapes into two sets of DVDs, one for each format. Then you are completely covered for the future. This Sony recorder also has an

Model: RDR-VX500
Manufacturer: Sony
URL: www.sony.com
Price: $399

iLINK (FireWire) connector so that you can connect your camcorder and record your movies to DVD easily. This combination unit also has a built-in television tuner and can play VHS tapes in addition to DVDs.

Slingbox

Have you ever been in a hotel room that doesn't get the same television stations you get at home? Or have you wanted to watch that big game in the garage where you can use your wireless Internet network, but you don't have a television? The Slingbox is your answer to either of these scenarios. Slingbox turns your Internet-connected laptop or desktop computer into your personal TV.

To use the Slingbox, you must have a broadband Internet connection and a router. The Slingbox is a small lightweight rectangular box that is hooked up to your television and router. The Slingbox makes it possible for you to access the programming from the connected television from any Internet-connected computer with a broadband connection. So if you hook up the Slingbox to your living room television, you can watch the big game on a computer in the garage, kitchen, or laundry room as long as it has a broadband Internet connection.

Model: Slingbox
Manufacturer: Sling Media
URL: www.slingmedia.com
Price: $249

With a Slingbox, you can actually watch your home television from anywhere in the world. The Slingbox might change your television viewing habits in unique ways. If you have a vacation home, you won't need to have a television set or cable or satellite TV service—just watch your home TV on your computer. The video quality is excellent. To control your home TV from afar, you are given a remote control on the computer screen. With that remote, you can do anything that you would normally do with your remote at home. You can change stations and even access programming from your TiVo or digital video recorder. This is one amazing little gadget!

Sandy's Lingo List

The world of technology has created some crazy new words. Here are explanations for a few of the more unusual words used in this chapter.

FireWire—A way to connect external devices that is also known as IEEE 1394. FireWire is faster than USB and supports data transfer rates of up to 400Mbps. Because of its high speed, FireWire is often used for transferring video files. FireWire was originally developed by Apple. It is found on most Apple computers and some PCs.

infrared (IR)—A wireless type of connection that works via infrared light waves. To use this type of connection, both devices must be equipped with infrared ports, be a few feet from each other, and have a clear line of sight between them.

liquid crystal display (LCD)—Used in display screens for most portable computers and many small digital devices. These displays have two sheets of polarizing material separated by a liquid crystal solution. An electrical current passing through the liquid causes the crystal to align, allowing or preventing light from passing through.

Media Center PC—This is a PC that uses Microsoft Windows Media Center Edition. It is a fully functioning desktop PC equipped with special hardware and software to display photographs and to display and record television and music.

media player—A generic term for devices that can play various types of electronic media. Media players are usually able to play both music and video.

MP3—The last three digits of a type of file called MPEG audio layer 3. This type of file is used for compression of audio signals. Because MP3 files are small, they can be transferred easily across the Internet and from device to device. This is the most popular file format for music devices, which are sometimes called MP3 players.

podcasts—Internet audio broadcasts that are presented in a series, somewhat akin to a radio or television series. You subscribe to a podcast and listen to it over the Internet or on your iPod or similar audio device.

radio frequency (RF)—Many wireless technologies are based on RF technology that sends radio frequency waves through the air.

satellite radio—Radio content distributed via satellite. Currently two companies, XM Radio and Sirius, provide satellite radio service. Each has its own programming, receivers, and satellites.

Chapter 4

Fun and Games

A hobby a day keeps the doldrums away.

—Phyllis McGinley

In your younger days, you probably danced to the Beach Boys as they sang "Fun, fun, fun, till your daddy took the T-bird away..." For most of us, those "fun" days of our youth were followed by a bunch of hard-working years. Now we are ready to have some fun again. Whether it's singing with friends, playing video games with the family, or using the computer and handheld games for a little enjoyment, we are ready to let technology lead us to some excitement and pleasure.

Video and Computer Games

Video and computer games just keep getting better. New technologies, faster processor chips, and better graphics make these games quite realistic. Listen to the crack of the bat in the baseball game, or feel the vibration of the wheel in the NASCAR game. It's easy to get into the "action" and have fun with these wonderful new digital diversions.

XaviX Baseball

The XaviX games are a new concept in gaming, a truly interactive television experience. The games respond to your physical movements, so when you're playing, you stand up, move around, and get a little exercise instead of just moving your thumbs.

Sandy's favorite

Setup is easy. You attach a small (10" × 10" × 4") device called the XaviXPORT to the standard audio/video ports on the television with the included cables. Insert the XaviX baseball game cartridge into the XaviXPORT. Choose the type of game, and you're ready to play ball.

You go through the motions of pitching without actually releasing the ball. (A wrist strap prevents accidental release.) What you see on the screen is your pitch put in play. You hear the crack of the bat and other realistic ballpark sounds, such as the crowds and the music. You determine which pitch to throw by pressing certain combinations of the four buttons on the ball. Onscreen prompts show you what buttons to press for the type of pitch you want, so you don't have to memorize button combinations. With the bat, you just stand in front of the television and swing and see the ball put

Model: XaviX Baseball
Manufacturer: XaviX
URL: www.xavix.com
Price: $49.99 for the game, $79.99 for the XaviXPORT

into play. Among options available in batting, you can choose to bat left-handed or to bunt. Both batting and pitching feel realistic.

The game includes a variety of players, stadiums, and games to play so that you don't become bored. XaviX also has tennis, golf, bowling, and bass fishing games.

Nintendo GameCube

Nintendo's GameCube platform has both good hardware and an excellent assortment of games. This game system is compact and just plain fun. The cube is 4.3" high, 5.9" wide, and 6.3" deep. Setup is easy. Plug the GameCube into your television, put the game in, and turn the cube on. An enclosed pamphlet and onscreen prompts get you started. The GameCube has a standard audio/video-out port and a digital-out port for HDTV.

The quality of the graphics is amazing. It's easy to mistake one of GameCube's football games for the real thing. Graphics move smoothly, and the sound is excel-

Model: Nintendo GameCube
Manufacturer: Nintendo
URL: www.nintendo.com
Price: $99.99

lent. The games come on small 3" disks, which you insert into the top of the console.

Nintendo has a wealth of games that play only on Nintendo systems. Titles include *Donkey Kong*, *Mario*, *Metroid*, *Kirby*, and *Pokèmon*, with something for everyone. There is plenty for grownups and kids to like. Some games, like *Kirby*, allow a five-year-old to play right alongside his father and grandfather, with all having a good time.

The GameCube has amazing expandability. It has four built-in controller ports, so you can easily plug in extra controllers. It can even accommodate wireless controllers. A game called *Mario Kart* had four of us playing and enjoying the game at the same time. The best thing about the

GameCube is that Nintendo offers plenty of games that the entire family can enjoy rather than the shoot-em-up games that are the mainstay of some other video game devices.

Donkey Konga

Here is an example of the kind of fun family games that you can purchase for the Nintendo GameCube. *Donkey Konga* is a musical adventure, rhythm action game. It works with a set of wood-like bongo drums that attach to the GameCube. Of course, Donkey Kong, the monkey, is

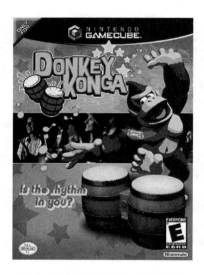

the main character of the game. His friends Diddy Kong and Cranky Kong make special appearances. It seems that Donkey and Diddy found a mysterious pair of barrels in the forest and realized they could be used as bongo drums. In *Donkey Konga*, you get to play the bongos.

The left side of the screen has a target. When you see the yellow icon line up with the target, you hit the left side of the bongos. When a red icon enters the target, you hit the right bongo. For a pink icon, you hit both bongos. When you see an icon that looks like a starburst, you clap. You do all this to the beat of the music. Songs include 30 easily identifiable ditties like "B-I-N-G-O," "Dancing in the Street," "Happy Birthday," "I've Been Working on the Railroad," "Itsy Bitsy Spider," and "The Loco-Motion."

Model: Donkey Konga
Manufacturer: Nintendo
URL: www.nintendo.com
Price: $49.95

Donkey Konga is deceptively simple. It seems easy enough when you start, but as you progress, it can be a real challenge. In advanced modes, it is an intense rhythm game. The themed backgrounds and different games keep you from getting bored. Each is a little different, like the 100m Vine Climb in the Banana Jungle where you harvest fruit to the beat of the music. There is also a street performance, a challenge, a battle, a jam session, and an ape arcade. If you are a hard rock fan, you won't like the music, but if you think it can be fun to sing "Row, Row, Row Your Boat," you'll love it.

At our house, three generations played *Donkey Konga* together. It's a real family game. Nintendo also has a game called *Donkey Kong Jungle Beat* that uses the bongos. It is an exciting action game that it not musical but uses the bongos in a completely unique way. It's not for older children and adults, but it has great graphics. You might want to check it out.

Atari Flashback Classic Game Control

Remember when you (or your children) played games like *Centipede*, *Breakout*, or *Asteroids*? Back in the 80s, when we played video games with a joystick and a couple of buttons, we thought we had stepped into a brave new world.

Now you can recapture the feeling of those days with the Atari Flashback Classic Game Console. It comes with the console containing 20 games, two Atari 7800-inspired joysticks, a power supply, and a TV cable. Two gamers can quickly and easily connect to the TV and start playing.

If you have a competitive youngster around who's been beating you consistently at current computer games, try challenging him to something

from the past. You can brush up your skills first with *Centipede*, *Asteroids*, *Millipede*, *Battlezone*, *Breakout*, *Air-Sea Battle*, *Warlords*, *Food Fight*, *Planet Smashers*, and more. The new games are great, but these early video games are still fun and entertaining.

Model: Atari Flashback Classic Game Control
Manufacturer: Atari
URL: www.atari.com
Price: $59.95

Also available are Atari Flashback2 with *3D Tic Tac Toe*, *Combat*, *Hangman*, *Video Chess*, *Haunted House*, and more Atari games for $29.99.

Ferrari Force Feedback Wheel

Given the current popularity of NASCAR and racing games, I thought a racing wheel to use with PC racing games would be fun. I was right. The Enzo Ferrari Force Feedback wheel is the exact replica of the wheel on the Enzo Ferrari, complete with the yellow prancing horse logo. It takes racing on the computer to a new level.

Although the directions are a bit sparse, the software is easy to install. The wheel mounts to the desktop with a clamp and hooks to the computer via a USB cable. For a more realistic experience, the foot pedals are included. You can use the wheel to play any PC racing games, but it shines when paired with a game that supports force feedback effects. This lets you feel the racing experience right through the wheel. The eight buttons and direction pad are built into the front of the wheel. Behind it are four paddles that function as gearshift levers and gas and brake levers. The wheel is comfortable and feels natural.

Model: Ferrari Force Feedback Wheel
Manufacturer: Guillemot
URL: www.thrustmaster.com
Price: $79

Unfortunately, this package doesn't come with racing software, but if you already have some or want to purchase some separately, the Ferrari Force Feedback Wheel is compatible with all PC racing games. You can buy a more expensive force feedback wheel, but this one offers good quality for the price and is a great place to start if this is your first driving wheel purchase.

Handheld Games

Do all those people you see playing with handheld games in airports and coffee shops know something you don't know? Or have you, like them, already discovered that these little devices can be fun? Whether you use them to whittle the time away while waiting for the doctor or you make gameplaying a part of your daily routine, handheld games hold a lot of excitement. The variety of these games is amazing. You can purchase an electronic device to play an updated version of an old game like Scrabble or Charades. Or you can buy the latest digital devices with touch screens and special effects. The choice is up to you. Maybe you need to try a few!

Nintendo DS

Sandy's favorite

Even if you are not a gamer, the handheld Nintendo DS (dual-screen) gaming system is an eye opener. The sturdy silver device is a clamshell design, which looks like an oversized PDA and opens to reveal two screens—one on top and one on bottom. The bottom screen is sensitive to touch, with two screens working in conjunction with each other. Objects and game characters can move from the top screen to the bottom, and you can control objects in the top screen by moving your finger or stylus across the bottom screen.

Buttons and controls are similar to other gaming devices but have additional unique functions. The Nintendo DS has wireless networking that allows you to send notes and doodles to other DS users and allows you to pit yourself against your opponents wirelessly. The DS is powerful and

Model: NTR S VKBA
Manufacturer: Nintendo
URL: www.nintendo.com
Price: $129.99

versatile. It has a rechargeable battery that lasts, on average, between 6 and 10 hours. It has full stereo sound, a calendar, and alarms. It also has a headphone jack and a microphone jack.

Each DS game comes in a hard plastic case that contains an instruction book and the game chip. A slot on the bottom of the unit can accommodate the larger Gameboy Advance (GBA) cartridges. Although the GBA games don't take advantage of the touch screen, they work flawlessly with the DS.

Although they haven't yet gotten around to creating games specifically for the boomers, many games are available for you to enjoy, including *Madden Football*, *Tiger Woods Golf*, *Texas Hold 'Em*, and *Super Mario*. At $129 plus the cost of the games, Nintendo DS is not cheap, but it is priced well for what you get. If you want to jump into the gaming world and look cool, get one for yourself.

New York Times Touch Screen Electronic Crossword

The New York Times Electronic Crossword is a lightweight handheld device about the size of a PDA. It has a small keyboard and a monochrome display and operates on two AAA batteries.

The selection of 1,000 crosswords from the *New York Times* newspaper offers a choice between cryptic and quick crosswords. You can scroll through the selections to choose the one you want.

The New York Times Electronic Crossword has some really cool features. It contains a crossword solver that finds missing letters, words, or whole puzzles with the tap of the stylus. It lets you save puzzles and come back to them later. I really like the feature that gives you hints.

Model: New York Times Touch Screen Electronic Crossword
Manufacturer: Excalibur
URL: www.excaliburelectronics.net
Price: $99.95

The screen is better than previous models, and it is lighted. However, you need good eyes for viewing the fairly small letters on the screen. The metal stylus and touch screen solve the problem of having the letter keys small and close together, but even so, you have to be dexterous to use it.

Electronic CATCH PHRASE Game

This is a party game in which two teams race against a timer. You can choose from six categories or choose an Everything setting that includes all six categories. Categories include Entertainment, Food, Places, The World, Sports & Games, and Sci-Tech.

It's best to seat the group in a circle and to alternate players so that each player is next to a member of the opposing team. The words or phrases appear on the CATCH PHRASE display. The player holding the CATCH PHRASE device gives clues to his teammates, who try to guess the word or phrase.

CATCH PHRASE is similar to the age-old game of Charades or the old *Pyramid* television show. In this version, the electronic device chooses the words and phrases, giving the game a lot of variety. It also adds an element of excitement because the gadget beeps as the game progresses. The beeping increases as time passes, putting pressure on the team that is trying to guess the word. Each team's turn ends when the phrase is guessed or the buzzer indicates that the time is up. One caveat: The beeping sound can be exciting, but some might find it aggravating.

Model: Electronic CATCH PHRASE Game
Manufacturer: Hasbro
URL: www.hasbro.com
Price: $24.99

This is one of those rare cases in which the old and the new meet in an entertaining electronic version of an old standard.

Hasbro Scrabble Express

This electronic handheld game from Milton Bradley is a portable version of Scrabble. It features four different Scrabble games, nine skill levels, a searchable 100,000 word dictionary, built-in scorekeeper, and letter shuffler. One player can use this battery-operated device to play against the computer, or two players can pit their wits against each other.

Scrabble Express has a color screen with adjustable contrast levels. Unfortunately, it is not backlit, so you can't play in a dark area. Although the lettering by the buttons is a little small, the buttons are easy to use and clearly labeled. You can turn off the sound if you'd prefer.

Model: 41301
Manufacturer: Hasbro
URL: www.scrabble.com
Price: $19.99

The game plays just like the Scrabble board game. It's good for practice and is also a fine learning tool. When you play against the computer, you can choose the computer's IQ level. Start with it at 1, and you won't feel bad if you are a beginner. As your skills improve, you can set the computer's IQ level higher and higher.

For those of you who enjoy Boggle, Hasbro also has a handheld Boggle game that is of similar size and is priced at $14.99.

The Official Scrabble Players Dictionary

This electronic dictionary does much more than the dictionary that is included in the electronic Scrabble game. This Official Dictionary is endorsed by the National Scrabble Association for recreational and school use and is authorized by the makers of SCRABBLE brand crossword games. This device instantly validates more than 100,000 words and provides brief definitions taken from the Merriam-Webster dictionary. It also has automatic spelling corrections and includes three Scrabble-type word games with adjustable skill levels.

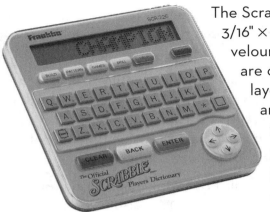

The Scrabble Dictionary measures a mere 4 3/16" × 4 13/16" × 5/8" and comes with its own velour pouch with drawstring. The letters are organized in the standard keyboard layout. Each key is inscribed with letters and the scrabble letter scores, giving the device an authentic look and feel.

The screen is easy to read, and the buttons are large enough to be workable for average-sized fingers. The included instructions are straightforward and easy to follow. Read them to get the most out of this device's functionality. Those of you who enjoy word games (crossword puzzles or otherwise) will enjoy this dictionary, and I could see myself growing dependent on it without much trouble. If you play Scrabble regularly, this could be a "must-have" for you.

Model: SCR-226
Manufacturer: Franklin Electronic Publishers
URL: www.franklin.com
Price: $59.95

Big Screen 20Q Deluxe

Can it read your mind? This small, portable game lets you try to stump the all-knowing artificial intelligence of the computer chip in this electronic device. It can be held in your hand or placed on a desk or tabletop.

To play, just think of any specific item and turn on the 20Q. Then watch in amazement as the some-

Model: Big Screen 20Q Deluxe
Manufacturer: Radica Games Unlimited
URL: www.radicagames.com/
Price: $19.99

what magical machine tries to guess what you're thinking! 20Q will ask you questions by scrolling them across the screen like a digital news ticker. You can answer Yes, No, Sometimes, Rarely, or Unknown to the questions. If you want to change an answer, you can use the Undo button. You can increase or decrease the speed of the text and even move backward and forward through it.

If the 20Q guesses the item you are thinking about within 20 questions, it wins. If not, you win! Part of the fun is that 20Q has an attitude. At times, it scrolls taunts across the screen like a little kid saying "Na, na, na... I'm going to win, and you're going to lose." It runs on two AA batteries (not included) and is recommended for ages 8 and up.

The 20Q intrigues all who play it. At this price, you can buy one for everyone you know! When I was trying it out, a 16-year-old and a 57-year-old were fighting to see which one could play next! It's sure to be a hit.

Maestro Travel Chess Computer

If you are looking for a game to keep your mind active, chess is perfect. But you might not always have an opponent available. With this hand-held chess game, you can play against the computer for a challenging bit of gaming anytime and anyplace.

Saitek makes an entire line of chess computers. This is one of its travel chess versions. It has 100 playing levels: 60 for fun, and 40 competition levels. You can save unfinished games. It also has a built-in chess clock. This device is for playing chess rather than learning to play chess. Yet, chess players are always learning to improve, and this device can help. It has 64 study positions. You can experiment and learn by taking back and replaying moves.

Model: Maestro Travel Chess Computer
Manufacturer: Saitek
URL: **www.saitek.com**
Price: $99

The Maestro Travel Chess Computer is a sturdy, well-made device with a blue LCD touch screen. It comes with a nice leather-like travel case. You use the included stylus to make your moves. The 2 1/2" × 3" screen is quite large for a handheld device, but because it must accommodate the entire chess board, each piece is somewhat small. The screen's backlight improves the screen's readability, but good eyeglasses or good vision is a must for using this device.

Saitek's computerized chess devices include additional travel games, several that are geared toward beginners, and several for intermediate users who would like to become experts. Garry Kasparov, the thirteenth World Chess Champion, endorses the Saitek products. Other Travel Chess versions are priced from $29.95.

Electronic Talking Texas Hold 'Em

When I was growing up, my dad and the male relatives played poker at most of the family gatherings. Sometimes the whole family would join in on a game of penny poker. They say everything old eventually becomes new again, and poker playing has proved this adage quite true. Poker is now wildly popular again. You can watch it on television, play with your friends, or take a trip to Vegas. You can also pick up a handheld electronic poker game.

This one uses the popular Texas Hold 'Em poker. Each table has six players. You play against five other computerized players. Unless you are proficient at this type of game, you need to read the directions as I did. After you get going, it's quite fun. This device

Model: Electronic Talking Texas Hold 'em
Manufacturer: Excalibur Electronics Inc.
URL: www.excaliburelectronics.net
Price: $29.99

even talks and announces the winning hand. Just like on TV, you can see your odds of winning with your first two cards. It's a fun game, but a well-lit place to play and good eyes are a must.

Just Plain Fun

We've already looked at video games, computer games, and handheld games. Now let's investigate the large variety of other fun items that technology has put in our grasp. You can have fun with musical toys, robots, and even electronic laser lights. And if you think all this fun is only for kids, think again! I've been playing with these high-tech toys and loving every minute of it!

Robosapien

Robosapien is a full-fledged, 12" tall robotic toy that mimics human actions. He walks, dances, whistles, and even gives you a high-five. You control his actions with an infrared light wave remote control similar to the one you use on your television. Robosapien can perform 67 different actions.

Although Robosapien can pick up light laundry and throw it in bags, he is more of a toy than a household helper. Children delight at his antics, and adults are engrossed by his large number of possible movements.

Robosapien 2 is 10" taller than his older brother and about $100 more expensive. He can interact with his surroundings. His new vision system allows him to wave when he sees you and reach out to shake your hand. He can detect

Model: Robosapien, original
Manufacturer: Wow Wee
URL: www.robosapienonline.com
Price: $99

obstacles, track movements, follow a laser light, and take objects that are handed to him.

Both Robosapiens seem to have an attitude that everyone finds appealing.

Electronic Dart Board

Because I'm not athletically inclined, even darts can be a challenge for me. But luckily, with the Electronic Dart Board, I can forget about my skill level and just have fun. The soft-tip darts are family-safe, taking this favorite bar game into the home.

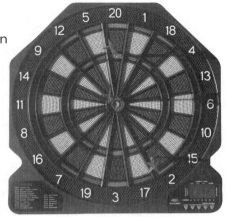

For one to eight players, the Dart Board keeps score electronically on its LCD screen. Its special lights, sound effects, and announcements will intrigue the whole family. It is a full-sized dart board that allows you to play 18 different games with 102 options. The battery-operated board comes with six nicely weighted soft-tip darts.

Model: Electronic Dart Board
Manufacturer: Excalibur Electronics Inc.
URL: www.excaliburelectronics.net
Price: $59.95

Bull's Eye! Throw a winning dart, and the sound effects of the board cheer you on. It's an enjoyable party game and wholesome family game.

Laserpod

If you liked the constantly changing flow and color of lava lamps, you will be fascinated by the Laserpod. It uses laser, LED, and crystal optics to create ever-changing light forms.

The base of the Laserpod has a large Swarovski crystal set into a glass refractor lens. This along with the lighting mechanisms are enclosed in a sleek aluminum tubular case that is about 4" high and 3" in diameter. When powered by the

Model: Laserpod
Manufacturer: Plugg
URL: www.laserpod.com
Price: $100

included AC adapter or three AA batteries, the Laserpod becomes a mini light extravaganza. The rotating crystal, blue LED lights and red laser beams produce an ever-changing display. The Laserpod comes with two electroplated domed diffusers. One is small and blue. The other is taller (about 6 1/2") and white. Placing each on the base produces a different effect. You can also let the base shine directly on the ceiling for an interesting effect. A small switch on the base allows you to turn the blue light on or off for even more variations.

Even with all these variations, I was left wanting the Laserpod to emit a few more colors. Perhaps future versions will do that. At $100, the Laserpod is pricey, but it is well-designed and constructed, and the light show is intriguing and relaxing.

OnKey Karaoke Pro

Sandy's favorite

Whether you have a good voice or you are tone deaf, this karaoke game can be fun. It comes with a sturdy microphone with a nice long 16' cable. Insert batteries or plug the unit into the wall and attach the cable to your television. Then choose the proper input channel on your TV, and you're ready for some fun.

The Pro model comes with almost 200 songs. Choose your song by looking up the song number on the included guide and pressing the number buttons on the microphone or by selecting it from an onscreen list of songs. The words appear on the television superimposed on different scenic backgrounds and images. If you choose, you can turn the background images off. Each word is highlighted as it is to be sung, and it's easy to follow.

The variations are almost endless. You can choose to turn the melody on or off. You also can control the tempo

Model: OKH08
Manufacturer: IVL Technologies Ltd.
URL: **www.onkeykaraoke.com**
Price: $199

and choose the key. Special effects change the sound from in a hall to in the shower or on a mountaintop with the press of a button. The voice-shifting feature can make your voice higher or lower or can change it from a man to a woman and vice versa. If you choose, you can even sound like an alien!

The feature list goes on and on. I really loved seeing the rating on the screen while I sang, letting me try to correct my singing on the go. I tried this karaoke player with several different groups of people, and they all loved it. Even those nonsinging "hold outs" were singing or humming along by the end of the evening. It comes with a nice carrying case. After you use it a few times, you will want to take it along and share the singing fun with others.

Singing Coach Unlimited

Perhaps you'd like to sing on Karaoke night, but you just don't feel comfortable with your singing voice. Here's the answer. Let the Singing Coach improve your singing.

Getting started is easy. Install the software, and plug in the included headset with microphone. Then set the record level and your vocal range. Next, start a lesson and start singing. The screen shows you the correct pitch and the pitch you are singing at. These visual cues are great, but it is not always easy to correct yourself. I advise going through the onscreen tutorial that shows you how to use the program. The tutorial also gives singing advice about good posture, belly breathing, and relaxation. The talking microphone in the tutorial is a bit hokey, but at least it's not boring.

Model: Singing Coach Unlimited
Manufacturer: Carry-a-Tune Technologies
URL: www.carry-a-tune.com
Price: $99.95

Expect to spend some time with this program. It has many impressive features, like allowing you to set the key, hear the beat, see a music view, and so on. You don't have to know anything about music to use the program; in fact, the program actually teaches you a little about music notation.

Twenty-four songs are included in the Singing Coach Unlimited. You can download 12 more of your choice from the Internet for free or purchase any tune from the manufacturer's large database of songs. You can even write your own music and lyrics if you are so inclined. This is a full-featured program, and it's cheaper than singing lessons.

Creative Prodikeys DM

The DM in Prodikeys DM stands for Desktop Music, but it could also stand for Desktop Magic. This keyboard allows you to have music on your desktop without having to add a dedicated music keyboard. It is a full-functioning computer keyboard and a musical keyboard at the same time.

The keyboard looks like a normal computer keyboard with a large palm rest. It has all the normal computer keys plus enhanced multimedia keys. It has Internet hot keys, programmable keys, and multimedia playback buttons (play, pause, volume, and so on). It operates just as you would expect when working on the computer, and it's comfortable with a good tactile feel.

Model: Creative Prodikeys DM
Manufacturer: Creative Labs
URL: www.prodikeys.com
Price: $99.99

The magic starts when you slide the cover off the palm rest and expose a 37-key midi keyboard. The Prodikeys comes with dedicated sustain, octave shift, and pitch buttons, in addition to a volume control wheel. To switch between the computer and the music keyboards, press the music key. To switch back, press the music key again. When you're in music mode, Prodikeys software appears on the screen, and the keyboard goes into music mode. In this mode, you can play the keyboard, create

music, or simply practice. The keyboard re-creates 128 different instru-
mental sounds, including piano, organ, drums, violin, and subcategories.
For instance, for organ, you can select Reed, Hammond, Percussive,
Rock, and Church. Choose the touch of the keyboard from Soft,
Normal, and Hard, the octave you are playing in, and the pitch. More
lessons and sounds are available from the Prodikeys website. Record
your creations to play back later. The better the sound card and speakers on your computer, the better the Prodikeys music will sound.
However, the instruments sound quite good with an average sound card
and ordinary computer speakers.

You don't really have to know anything about music to enjoy playing
with the Prodikeys keyboard. A professional musician would probably
find it boring, but children and adults who are interested in music will
find it fascinating.

AIBO

The name AIBO stands for Artificial Intelligence roBOt, but AIBO is also
named after the Japanese word for *pal*. AIBO looks like a robotic dog,
but he is meant to be a friendly buddy, a constant companion for any
lonely or not-so-lonely human.

AIBO is outfitted with cameras and
sensors that make him understand and
respond to humans. You raise an AIBO
puppy to adulthood, just as a real
dog. AIBO puppies can act just like
mischievous little puppies, each
with its own personality, often
refusing to obey. Yet they slowly
learn. They thrive on interaction
with their owner. Just as with a real
pet, you have to spend a lot of time
with these robotic pets. As you invest
more time in them, AIBOs develop
personalities and become more fun.

Model: ERS-7M2
Manufacturer: Sony
URL: **http://www.sony.net/Products/
aibo/**
Price: $1,999.99

The amazing AIBO is capable of many functions. He can take pictures, guard the house, or even wake you up in the morning. He also has emotions. An AIBO can be happy, sad, afraid, or surprised. He wags his tail, moves his silicon ears, and has a screen-like face that conveys his feelings to his owner. He can play with his pink ball or carry his bone in his mouth.

Specially imprinted AIBO cards can be used to issue commands like sit or dance. When AIBO senses that his battery is low, he seeks out his recharging unit. When he is completely recharged, he wakes up ready to play and interact with his owner. Every version of the AIBO dog gets better and better. The first AIBOs were just cute toys, but the newer AIBOs are beginning to feel like real companions.

Sandy's Lingo List

The world of technology has created some crazy new words. Here are explanations for a few of the more unusual words used in this chapter.

crystal optics—The multifaceted properties of crystals that make light behave in interesting ways.

handheld—This term refers to compact items that you can operate while holding in your hand. Technology has allowed us to create smaller and smaller devices, creating handheld devices from items that were previously much larger.

karaoke—The singing of popular songs to prerecorded music.

laser—A common term referring to a concentrated light technology. The name is derived from **L**ight **A**mplification by the **S**timulated **E**mission of **R**adiation. Laser technology is used in everything from toys to telephone systems.

liquid crystal display (LCD)—Used in display screens for most portable computers and many small digital devices. These displays have two sheets of polarizing material separated by a liquid crystal solution. An electrical current passing through the liquid causes the crystal to align, allowing or preventing light from passing through.

wireless—A generic term that refers to any nonwired transmission of data. Although this includes radio, television, and cell phones, it is most often used to refer to nonwired networked computers and devices, such as keyboards and mice that are not attached to the computer (or other device) by wires.

Chapter 5

Vision Aides

I thank God for my handicaps, for through them, I have found myself, my work, and my God.

—Helen Keller

Nearly everyone over age 45 experiences *presbyopia*, a condition that adversely affects up-close vision and causes people to reach for the reading glasses. As the eyes age, they become more susceptible to other diseases, too. Almost 40 million people in the United States suffer from cataracts. At age 55, only 15% of people are affected. By age 85, 90% of us have cataracts. Macular degeneration and many other eye problems are also known to be age related. The good news is that no matter what vision problem you suffer from, a multitude of gadgets and tools is available to aid you.

Magnifiers and Large Text

Remember the song "I can see clearly now"? I always think of that song when I struggle to see tiny print or when my eyes tire from using the computer.

Many of us use glasses or contact lenses to help us see clearly. When glasses aren't enough, we have many other options. This section of the chapter details some of the good old-fashioned gadgets and the high-tech options to help you see clearly.

Max Portable Magnifier

Max is a lightweight, portable digital magnifier. It hooks up to any TV either by the included RCA or the RF cables. Because Max is only slightly larger than a computer mouse, it is easy to place it on the book, newspaper, label, or other object you want to magnify. The full-color magnified image appears on the television screen.

Max magnifies from 16X to 28X on a 20-inch TV. It can also display in high contrast black on white or white on black, depending on your vision needs. Max even works on curved surfaces for use on cans, bottles, and medicine containers. Several versions of Max are available, including one with an integrated screen.

Manufacturer: Enhanced Vision
Distributor: Gold Violin
URL: www.enhancedvision.com
Price: $449

QuickLook Magnifier

This is one high-tech magnifier! QuickLook uses digital magnifying technology that was previously available only in large cumbersome devices, so it is really cutting edge. Measuring 6 1/2 inches by 3 3/4 inches by 1 inch and weighing about 10 ounces, QuickLook is extremely portable. The 4-inch lighted color screen makes it easy to read everything from medicine bottles at home to menus in a restaurant. It has 6X magnification that seems stronger because it is displayed on a large reading screen. This magnifier has six levels of brightness, and you can change the display from color to a black and white or reverse image with the touch of a button.

QuickLook has a lot going for it. It has an integrated camera that can be rotated easily, the capability of varying the magnification by changing the viewing angle, and a carrying case with belt clip and rechargeable battery. And the clarity is amazing! The $795 price tag is its only drawback.

Model: QuickLook Magnifier
Manufacturer: Ash Technologies
URL: www.quicklook.com
Price: $795

Big Eye Magnifier

If you are looking for something that is much more affordable, try the Big Eye Magnifier. This small magnifier is made of acrylic and shaped like a camera lens. It is available in both 5X and 7X versions. It is 3 to 4 inches in diameter and 2 1/2 to 3 inches tall. The 5X lens is 2 1/2 inches in diameter, whereas the 7X lens is 1 1/2 inches in diameter. You can obviously see more text in the 5X lens, but the magnification in the 7X lens makes items appear larger.

You slide the Big Eye over the item you want to see more clearly. Although it's not very high-tech, this magnifier is helpful when you need to increase the size of any text or small object.

Model: 7X Big Eye Magnifier
Manufacturer: Gold Violin
URL: www.goldviolin.com
Price: $15 for 5X, $10 for 7X

2-in-One Lighted Magnifier

Yes, we return to the days of yesteryear. This looks like the round magnifier with a handle that Sherlock Holmes used to find clues, yet it is updated with some useful features. You can set the handle into a bendable base, which turns it into a hands-free unit. This magnifier has a built-in light for viewing in dark or dimly lit circumstances. The acrylic lens provides 2X magnification and is 4 inches in diameter. It also has a small bifocal lens near the base that magnifies 4X. It has nice heft and a good feel.

Model: 2-in-One Lighted Magnifier
Manufacturer: Gold Violin
URL: www.goldviolin.com
Price: $22

Reizen 15X Illuminated Pocket Magnifier

A simple product that can fit in your pocket, this sturdy little magnifier gives 15X magnification. When you need to make something "big," just pull this out of your pocket. A press of the button illuminates the 28mm diameter lens for use in dark and dimly lit situations. It comes with a small protective pouch and two AA batteries.

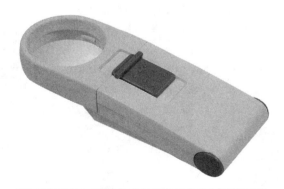

Model: Reizen 15X Illuminated Pocket Magnifier
Manufacturer: Reizen
Distributor: Maxi-Aids
URL: www.maxiaids.com
Price: $24.95

Royal Talking Watch

The 1 1/4-inch dial, clear black numbers, and heavy black hour and minute hands on a white face make this watch easy to read. Even better, with the press of a button, the watch announces the time and date in a clear male voice. The watch is powered by a quartz movement and two included lithium batteries. It is also a talking alarm clock with a built-in timer. On top of that, it is good-looking enough to be taken as an everyday functional fashion watch.

Model: Ladies Tel-Time Bi-Color with Expansion Band
Manufacturer: Royal
Distributor: Maxi-Aids
URL: www.maxiaids.com
Price: $49.95

Lighting

Have you ever struggled to read some tiny text and clarified your vision by holding the text under a light or near a window? Shedding a little light on whatever you are trying to view can help you see clearly. Today you can buy special lamps for almost every situation.

Natural sunlight is one of the best light sources for our vision. It allows us to see with remarkable clarity and vibrancy. Yet, obviously it is not always available. When Edison invented the light bulb, he never envisioned that light bulbs would be created to mimic natural sunlight. Yet "sunlight" lamps are readily available today and can be a blessing for those who have aging eyes or vision problems.

OTT-LITE VisionSaver™ Plus

Sandy's favorite

This lamp looks good and provides you with a good look at whatever you are viewing. The brushed nickel finish and sleek design appear at home in any environment. The 24-inch lighted arm pivots up and down, making it perfect for reading and detailed hobby work and working on the computer.

The Vero table lamp is part of Ott-Lite's VisionSaver line. The patented technology replicates the wavelengths in the visible spectrum of sunlight in balanced proportions. It uses a specially formulated blend of rare earth phosphors to create an artificial light source that looks and feels like natural daylight. Ott-Lite not only improves your visual acuity, but it has no glare to stress your eyes. It also produces TrueColor lights, which are specifically aimed at crafters,

Model: VisionSaver™ Plus Vero Table Lamp T59BNR
Manufacturer: Ott-Lite
URL: www.ottlite.com
Price: $89.95

and Natural Light Supplement light, which is perfect for plants and is said to help counteract daylight deprivation disorders.

Intelli-Lite

The Intelli-Lite views sunshine as powerful medicine. We all need sunshine to survive. The Intelli-Lite has two movable lighted arms and a center light. All are created to mimic natural sunlight. Like other natural sunlight lamps, the Intelli-Lite provides clear light for crisper vision. Because it was also developed for daylight deprivation problems, some physicians have been known to write a prescription for this light, making it a tax deduction.

> **Model:** Intelli-Lite Bio Tasklight
> **Manufacturer:** Science of Light, Inc.
> **URL: www.bioenlightenment.com**
> **Price: $299.95**

Feather-Touch Keyboard Light

Sometimes you just need to shed a little light directly at your subject. The Feather-Touch Keyboard Light does just that. Sometimes lowering the room lights provides a nice contrast for the computer monitor, but you still need to be able to see the keyboard. That's when this light is perfect. It is light and slim. The base is a small silver pad that appropriately looks just like a computer mouse. Press the mouse buttons to turn the light on and off. A slim 10-inch silver arm extends over the keyboard to illuminate the keys. The cathode fluorescent light remains cool and is rated for 20,000 hours of use.

> **Model:** Feather-Touch Keyboard Light
> **Manufacturer:** Gold Violin
> **URL: www.goldviolin.com**
> **Price: $28**

Verilux Happy Eyes

The EasyFlex gooseneck allows the head of this lamp to rotate freely so that you can position it right where you want it. HappyEyes lamps are also designed to mimic natural sunlight. The light that emits from the HappyEyes lamps is much whiter and not as natural-feeling as the Ott-Lite or the Intelli-Lite. Yet, the natural white light can reduce eye-strain and glare and heighten visual acuity. A utility tray that attaches to the lamp stem is available for holding small items that you might be working with. If you are interested in a natural light lamp, it is worthwhile to compare lamps from various manufacturers because they use different technologies to produce their lights.

Model: HappyEyes® EasyFlex Floor Lamp
Manufacturer: Verilux
URL: www.verilux.net
Price: $129.95

Long-Life 3X Magnifying Floor Lamp

This attractive floor lamp is available in brass or satin nickel finish. You move the lamp's arms to position the 3X magnifying lens over the area you want to focus on. The 3-inch ground glass lens magnifies, and the light helps you to see. The magnifying floor lamp has a cathode fluorescent bulb that lasts more than 10,000 hours and uses only five watts of electricity.

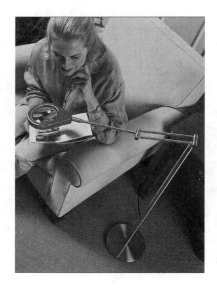

Model: Long-Life 3X Magnifying Floor Lamp
Manufacturer: Gold Violin
URL: www.goldviolin.com
Price: $69.96

Specialty Glasses

For the past two centuries, we have relied on eyeglasses to correct our vision. Put on the correct pair of glasses, and we can happily start singing that "I Can See Clearly" song. Glasses can correct nearsightedness, farsightedness, and astigmatism. New high-tech glasses can also help various eye problems and can even help the blind see.

JORDY Glasses

Remember Geordie LaForge, the blind engineer on *Star Trek*? He wore a visor over his eyes that enabled him to see. Although it sounds futuristic, this type of technology is here now. The aptly named JORDY 2 glasses look futuristic. They incorporate a miniature, auto-focus digital video camera to project images directly onto the user's eye at up to 30x magnification. This allows many with low vision conditions to see again. Wearing JORDY glasses, even a highly vision-impaired person can read books, watch television, and see objects in the room.

Product image not available for reproduction.

See vendor website.

The JORDY 2 headworn display works by using two CCD digital cameras that are controlled by a miniature computer that changes contrast magnification and brightness and replays the image on two postage-size digital television

Model: JORDY 2
Manufacturer: Enhanced Vision
URL: **www.enhancedvision.com**
Price: **$2,795**

sets, all built in a goggle-type device that weighs less than 8 oz. The JORDY is paired with a desktop plug-n-play docking station for additional flexibility to read and write in stationary environments. At almost $3,000, these glasses are pricey, but they are valuable for many and have other unique features.

SOLA Access Computer Glasses

Many bifocal lens wearers have difficulty seeing the computer screen. They move their head up to see through the bottom of the bifocal lens and then move their head down to see through the top. Often, they resort to removing their glasses completely. None of these viewing methods works well, leaving the computer bifocal wearer frustrated. Bifocals correct both near and far vision, but they don't address problems with the mid-range vision that is required for computer viewing. There is, however, a solution: Special lenses like the SOLA Access Lenses are progressive lenses specially designed for clear uninterrupted vision with distances up to 7 feet. They are perfect for working on the computer. Besides computer users, SOLA Lenses are popular with artists, musicians, and hobbyists.

Model: SOLA Access Computer Glasses
Manufacturer: SOLA
URL: www.sola.com
Price: Varies

Your eye doctor must prescribe these lenses. Similar lenses are made by several other companies.

FlipperPort with Glasses

The FlipperPort is a high-resolution rotatable color camera and a pair of glasses that display the magnified image. The port is portable, so no matter where you go, you can simply point the camera at what you want and see the resulting image in the lightweight glasses. The FlipperPort is an entire image enhancement system with a lot of flexibility. It can also connect to a TV or computer monitor. For reading text, you

Model: FlipperPort with Glasses
Manufacturer: Enhanced Vision
URL: **www.enhancedvision.com**
Price: $2,395

can purchase a FlipperPanel with a 7-inch or 10-inch LCD screen where the image can be projected. The FlipperPort allows 6X to 50X magnification.

MelaOptix Glasses

Melanin glasses come in two lens colors. The pale yellow is used for computer work and night driving. The dark brown is for wearing in the sun. These glasses are unique in that melanin is extruded into the lens material when the glasses are made. Melanin reduces blue light, which in turn reduces glare and enhances vision.

Sandy's favorite

Not surprising, just as we lose our dexterity and our vision as we age, we also lose melanin. By age 50, about 25% of our melanin is lost. Many believe that melanin glasses provide a way to compensate for lost melanin.

I can tell you that I love my melanin glasses. They reduce glare and keep my eyes from straining in the sun and at the computer. Prescription MelaOptix glasses are also available.

Model: MelaOptix sunglasses and computer glasses
Manufacturer: MelaOptix
URL: **www.melaninvisioncenter.com**
Price: $79

Vision Software

By some definitions, software is not a gadget or a gizmo, but when the software helps you see the computer screen better or helps you to interact with the computer more easily, it certainly can be considered a useful gadget.

WebEyes

Sometimes you need a little relief from the eyestrain that Web browsing causes. Or you just need to make that text a little larger to make your surfing easier. That's

where WebEyes excels. You can choose text sized from 4 to 144 points manually from a drop-down box on the WebEyes toolbar. The enlarged text is smooth, with no jagged edges.

WebEyes has an excellent feature called Read Like a Book. When you click on the Book icon, a new window opens. The same web page appears, but it is formatted in several columns and looks somewhat like a book. When you simply enlarge the type on a web page, you often have to scroll through a lot of enlarged text. The Read Like a Book feature eliminates all that scrolling by presenting you with a nicely formatted page. This feature also gives you the option of viewing text only, text with illustrations, or text with markings where illustrations would be on the website. Excellent instructions and an interactive demo make it easy to use WebEyes. Unfortunately for Mac users, WebEyes is a Windows-only program.

Model: WebEyes
Manufacturer: Ion Systems, Inc.
URL: www.ionwebeyes.com
Price: $24.95

ViaVoice

You talk. It types. This software is useful for those who have vision problems, but it is also popular with those who don't like to type or who want to dictate into a portable recorder and have their words automatically typed into the computer. ViaVoice works with Microsoft Word, Excel, and several other programs. In Word, Excel, and Internet Explorer, you can also give voice commands like "open" and "save." Speak into Outlook Express, Outlook, and some other e-mail programs, and ViaVoice types your message for you. This version includes a noise-canceling USB headset with an attached microphone. You must

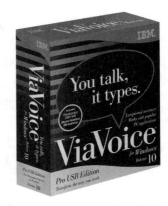

Model: IBM ViaVoice Pro USB Edition Release 10
Manufacturer: ScanSoft
URL: www.scansoft.com/viavoice
Price: $189.99

spend some time training the computer to interpret your voice correctly, but the results are worthwhile. You can talk and have the computer be your own personal typist. A similar version of ViaVoice is available for Mac OS X users.

ZoomText

The ZoomText software magnifies the screen up to 16 times. The problem with many screen magnification programs is that the magnified images are grainy and the text is choppy. ZoomText's smoothing technology makes sure that the magnified images are clear and the text is sharp. Special pointer and cursor enhancements make it easier to find the cursor on the screen, and the Web link finder makes it easier to find links. This software also lets you change the screen colors for improved contrast.

Model: ZoomText 8.1 Magnifier/ScreenReader
Manufacturer: Ai Squared
URL: www.aisquared.com
Price: $595

This program really shines for text readability. The Line Zoom window magnifies a single line at a time while automatically adjusting the height of the other text in the document. This program also reads text to you. With Windows XP, even PDF documents can be read aloud. ZoomText is a fully functioning magnification and screen reader program with a lot of useful functionality. Unfortunately, no Mac version is available.

BigShot

BigShot does only one thing, but it does that thing well. It magnifies the screen. It works with all Windows programs. BigShot has an onscreen toolbar that can be used to choose the level of magnification from 100% to 200% in increments of 5%. You can magnify the entire screen or only the active window. When everything on the screen is larger, you naturally have to scroll up and down or side to side to see all portions of the screen. BigShot actually does this for you. If you move your cursor to the top, bottom, or side of the screen, the unseen portions of the screen automatically scroll into view. This program is a good option if you need to give the screen size a little boost to ease the strain on your eyes. BigShot is a Windows-only program.

Model: BigShot
Manufacturer: Ai Squared
URL: www.aisquared.com
Price: $99

JAWS

JAWS is a fully functional screen reader for the blind. It is the most popular program of this type in the world. It has an internal speech synthesizer that reads aloud information from the screen. JAWS also outputs to Braille displays, providing the most Braille support of any software program. It is available in 17 different languages and comes with five hours of

Model: JAWS Standard
Manufacturer: Freedom Scientific
URL: **www.freedomscientific.com**
Price: $895

audio training. JAWS works with all standard Windows applications and includes advanced support for popular applications like Microsoft Office. It has extensive functionality and even includes a unique scripting language for further customization with non-standard Windows applications. JAWS is a Windows-only program.

Sandy's Lingo List

The world of technology has created some crazy new words. Here are explanations for a few of the more unusual words used in this chapter.

Computer Vision Syndrome (CVS)—The American Optometric Association defines CVS as eye and vision problems related to computer use. The symptoms of CVS include eyestrain, periodic blurred near vision, headaches, and dry, tired, sore, red, or burning eyes.

Daylight Deprivation Disorder—This disorder is caused by lack of sunlight, as is another similar disorder called Seasonal Affective Disorder (SAD). These disorders cause moodiness, social withdrawal, and lack of energy.

magnification—When the power of a magnifier is described as 5X or 10X, it means that the image viewed through the magnifier is 5 or 10 times larger than the same image viewed through the human eye.

melanin—A pigment that occurs naturally in your body. It gives your skin and hair its color, and, in the eyes, it protects against damaging light rays by absorbing light over a broad spectrum of ranges.

Chapter 6

Accessories for Healthy Living

The march of invention has clothed mankind with powers of which a century ago the boldest imagination could not have dreamt.

—Henry George

They often say that the three most important things in life are health, wealth, and happiness. Of course, health is the first of these because without it, we cannot enjoy the rest of our lives. As we age, we gain a certain amount of wisdom. With this wisdom comes the realization that we must do everything we can to stay healthy.

Scientists and researchers have given us information that our parents and grandparents didn't have. We now know that smoking is harmful to our health and that too much sun exposure can be dangerous. We understand the importance of nutrition and exercise. Technology has also come to our aid with plenty of gadgets that can help us in our quest to retain our health.

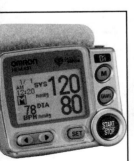

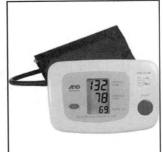

Staying Healthy

Youngsters often take their health for granted. Those of us who have lived more than three or four decades quickly realize how important health is. We strive to live a healthy lifestyle and try to protect our health in any way we can. Today's gadgets and gizmos make living healthier easier and more fun.

Coolibar Sun Protective Clothing

Sandy's favorite

One hundred years ago, people covered their bodies to accommodate their standards of modesty. Women wore long skirts. Men wore long sleeves. No one left the house without a hat. As our standards changed, we shed our clothes, never once considering that the sun could be damaging to our bodies. Now we know better. Experts tell us that too much sun can cause skin cancer, one of the most common types of cancer in the world.

Coolibar uses various high-tech methods to build sun protection right into their clothing. The women's everyday blouse that I tested was made of tightly woven microfibers that are enhanced with sun-protective titanium dioxide. The hang tag on the blouse assured me that it blocked 97% of the damaging UV rays from the sun. Other products, such as the protective swimwear, are made of tightly knitted synthetics that are resistant to damage by chlorine and salt water. All the Coolibar products that I looked at were well made yet soft and comfortable.

Model: Everyday Blouse with Sun Protection
Manufacturer: Coolibar
URL: www.coolibar.com
Price: $59.95

Coolibar uses extensive testing to guarantee the sun protection of their garments, which remains with the product even after repeated washings. Coolibar products have received the Skin Cancer Foundation Seal of Recommendation. Coolibar also makes other sun-protective clothing including shirts, hats, pants, and swimwear for men, women, and children.

Digitt UV Monitor

Sometimes the sun can be stronger than you realize. I recently had a terrible rash on my neck and chest. When the doctor said that it was a reaction from the sun, I protested that I hadn't been out in the sun. We determined that the sunlight shining at me directly through my office window was the culprit (along with some new medication that I was taking). Sometimes you can be exposed to the damaging rays of the sun without even realizing it. So if you are sun-sensitive or you want to make sure you limit your sun exposure, it might be helpful to be able to monitor the amount of ultraviolet sun in any given situation. Here is the little gadget that can do just that.

The UV Monitor is about 1 3/4 inches × 3 inches and has a small curved arm that allows it to be hung just about anywhere. Although the LCD screen is small, it is clear and easy to read and has a built-in backlight. Functionality is controlled with five buttons. Pressing the mode button lets you scroll through four screens that include the time and date, temperature, UV index detail, and a sunbathe timer that auto detects the UV index every 10 minutes. It even has an alert when the UV index rises above 7.0. You might need to get out the magnifying glass to read the print on the instructions, and the monitor is not the sturdiest gadget I've ever seen, but all in all, it gets the job done.

Model: Digitt UV Monitor
Manufacturer: DMD Design Manufacture Distribution, LLC
URL: www.alldmd.com
Price: $11

Heart Rate Monitor

If you have been sedentary for most of your life, it's not too late to get into an exercise program. Studies show that to get the most benefit from an exercise program, you need to exercise at the right intensity. You want to push yourself a little, but not push too hard. A heart rate monitor is a way to accurately measure your exertion level to get the most out of your exercise time.

Like most heart rate monitors, the Polar FS3 comes with two parts. The transmitter is attached to a soft mesh band that is worn around the chest. The receiver looks and is worn like a wristwatch. Although some monitors have three, four, or five

Model: FS3
Manufacturer: Polar
URL: www.polarusa.com
Price: $69

control buttons, this model has only one. In its default setting, the watch face displays the time and date in large clear numbers. Press the button to switch to the exercise mode, and the timer automatically starts. When you are wearing the transmitter and the monitor is in the exercise mode, your heart rate is displayed on the screen along with the amount of time that you have been exercising. Press the button to stop the exercise time. Press again to set the time and date and to input your age and target heart rate.

Although I thought the one-button scheme would be difficult to maneuver, I actually found it to be simple to navigate. The large letters and clear screen of the FS3 are a dream, and the black rubber watch looks quite good. Wearing it makes you feel like you're an exercise guru even if you're not!

Pedometer with Panic Alarm

No one really needs a pedometer, but it can be an incentive to exercise. If you're anything like me, you need plenty of motivation when it comes to exercise. Simply put, a pedometer keeps track of the distance that you travel by registering the number of steps taken. You set the length of your stride and your body weight. Then clip the pedometer to your belt. Because it's small and lightweight, you barely know it's there.

A mode button steps you through three screens where you can see the steps taken, calories burned, time elapsed, and total distance traveled. This little gadget also displays the time so it can double as a watch when you are walking. It has yet one more

> **Model:** Pedometer with Panic Alarm
> **Manufacturer:** Oregon Scientific
> **URL: www.oregonscientific.com**
> **Price: $29.95**

feature that I just love. The unit has a small cord with a clip. You can use this clip to attach to your clothing, but more importantly, if you pull out the cord, it sounds an alarm that is loud and clear. I almost jumped out of my skin the first time I tried it. If you encounter a medical emergency or an unexpected danger when you are walking or jogging, this device could be a life-saver. The alarm is quite secure and not easily set off accidentally. Yet it would be simple to pull the cord in case of emergency. The dual purpose of this device makes it a winner.

Talking Pedometer

Clip it on your belt and start walking. The Talking Pedometer displays on the screen the number of steps and the distance traveled. Press the Talk button, and this device verbally announces the number of steps and distance. A talking alarm can tell you when you have reached your distance goal.

> **Model:** Talking Pedometer
> **Manufacturer:** Maxi-Aids
> **URL: www.maxiaids.com**
> **Price: $7.25**

The Talking Pedometer also plays seven melodies with different tempos. It won't win prizes for sound quality, but it can help you synchronize your walking speed and make walking a bit more fun.

Talking Bathroom Scale

If you are dieting and you dread getting on the scale every week and seeing that same too-high number, this scale gives you a change of pace. It uses a pleasant feminine voice to tell you how much you weigh. If you tire of the voice, you can turn it off and view your weight on the 1-inch digital LED display.

The talking scale also comes in handy if you have a vision problem or simply don't want to have to put your glasses on to view the scale. It's one smart scale. It speaks English, French, and German and can weigh in pounds, kilograms, or stone.

> **Model:** Talking Bathroom Scale
> **Manufacturer:** Gold Violin
> **URL: www.goldviolin.com**
> **Price: $69.95**

LifeSource® ProFIT Personal Health Scale

Keeping your weight in control is a large part of the healthy life that we all seek. This gadget doesn't help you lose weight, but it does keep track of your weight and can even be an incentive for losing weight.

First and foremost, it is a precision scale, accurate to the tenth of a pound. At less than 5 pounds, the scale itself is kind of "scaled down." At a little more than 12 square inches, the gray case has

> **Model:** UC-321
> **Manufacturer:** A&D Medical/LifeSource®
> **URL: www.lifesourceonline.com**
> **Price: $150**

a thin profile and attractive simplistic look. The numbers on the LCD screen are large and clear.

More importantly, this scale can do more than just weigh you. If you enter your target weight, the scale tells you how much further you have to go to reach your ideal weight. It is at least a slight variation of having that same number stare you in the face each morning. If you input your height, you can also use the scale to measure your BMI. In addition, the scale has a built-in memory function to track your weight history. All this might be more information than you want to be faced with, but it can really be an incentive to keep on track.

The only drawback is that the switch to change to the BMI, Memory, and Target weight is on the bottom of the scale. So you have to lift it up every time you want to switch modes. Perhaps that is part of LifeSource's effort to help us keep fit.

XaviX Health & Fitness Manager

The XaviX Health & Fitness Manager represents a new way to manage your family's fitness. This product includes a special scale that looks like the average scale with a sleek design. It also comes with a cartridge that is inserted into the XaviXPORT (purchased separately). The scale wirelessly transfers your weight to the XaviXPORT, which looks like a small game console and is attached to the television.

Your weight appears on both the LCD display on the scale and on the television. The XaviX fitness system also tracks other information, including your BMI. In fact, it can track information for up to four family members for a full year. Also included is a unique Health

Product image not available for reproduction.

See vendor website.

Model: XaviX Health & Fitness Manager
Manufacturer: XaviX
URL: www.xavix.com
Price: XaviXScale $99.95, XavixPORT $79.99

Management Diary that can be used to track progress. The Family Fitness Manager has 10 well-being programs with guidance for a healthy diet, basic yoga techniques, and words of encouragement. With this product, your television can help you manage your health instead of turning you into a couch potato.

You can also use the XaviXPORT for interactive games like baseball, tennis, golf, bowling, and fishing.

Weigh of Life Nutritional Scale

We grew up on Big Macs and French fries. It's not surprising that all that unhealthy eating has manifested itself in flabby bodies. As we age, healthy living often starts with a change in eating habits. The Weigh of Life Nutritional Scale is just the tool you need to monitor your food choices and portion sizes. It can also be valuable if you need to track your sugar, cholesterol, or sodium input.

Model: Weigh of Life Nutritional Scale
Manufacturer: DMD - Design Manufacture Distribution, LLC
URL: www.alldmd.com
Price: $35

You can use the Weigh of Life for simple portion control, but it does much, much more. This scale measures 11 key nutritional elements including the three previously mentioned plus calories, protein, bread equivalent, carbohydrates, fiber, fat, saturated fat, and fiber. Food is entered on the keypad in the front of the scale by the code number. The codes are provided on the small but readable plastic-coated pamphlet. The weight results along with all the nutritional information for the item weighed are provided on an ample-sized LCD screen.

You need to spend some time learning the intricacies of this scale, but basic functions are easy after you get the hang of it. Oh, I should mention that the stainless-steel look and clear glass dish make this scale an attractive addition to any kitchen.

Personal Health Care Devices

When I was growing up, a mercury thermometer was the only personal health care device in most homes. Times sure have changed. Today you can monitor your blood pressure and test your cholesterol at home. Even thermometers have gone high-tech. No more mercury—and now thermometers can even talk to you.

Lifestream Cholesterol Monitor

According to the American Heart Association, more than 100 million American adults have a total cholesterol level of 200 mg or higher that should be monitored. More than 40 million have levels above 240 (which is considered high risk). This FDA-approved Cholesterol Monitor helps you keep track of your levels without a trip to the doctor's office.

The device uses test strips that come with a calibration key, lancet, alcohol swab, and Band-Aid. A handy storage area under the device can be used to hold the test strips and miscellaneous paraphernalia. Enter the provided calibration key into the device. Then put the lancet into the included lancing device, swab your finger, and place the lance on your finger. Press the button, and the lance pricks your finger. (Ouch!) Insert the testing strip into the device, placing a large drop of blood on the indicated place on the strip. After several minutes of processing, the LCD screen shows the cholesterol level in your blood. Only the total cholesterol level is given, with no breakdown. Although the process is easy, I had several misses before I got the hang of it, wasting several of the pricey test strips. Yet, if you need to track your cholesterol, this device can do it for you.

Model: Lifestream Cholesterol Monitor
Manufacturer: Lifestream Technologies, Inc.
URL: www.lifestreamtech.com
Price: $119.95, plus $19.95 for 6 test strips

LifeSource® Blood Pressure Monitor

High blood pressure, or *hypertension* as it is called, is rampant in our society. Although you can run over to the doctor's office or drug store to get your blood pressure measured, having a blood pressure monitor at home makes it easier to track and control your blood pressure. The LifeSource One Step Plus Memory is fully automatic. The arm band inflates to the correct pressure each time a measurement is taken. It also stores the previous 30 readings and averages the readings for you. Wrap the cuff around your arm and press one button to measure your blood pressure and your pulse rate. The result is shown in large numbers on the LCD screen.

Model: UA 767PAC One Step Plus Memory
Manufacturer: A&D Medical/ LifeSource®
URL: www.lifesourceonline.com
Price: $99.99, medium and large cuff

This device can run on four AA batteries or the included AC adapter. The included quick start guide and instruction manual are clear and well written, giving specific instructions for getting an accurate reading.

Besides the built-in memory, this monitor has two other impressive features. First is its ability to provide blood pressure and pulse rate readings even when an irregular heartbeat occurs. Second is the fact that it is clinically validated for accuracy. The only drawback of this device is that to take advantage of the memory feature, only one person can use the unit; however, LifeSource does have a model two or more people can use (UA-774AC).

Omron Wrist Blood Pressure Monitor

This is a wrist blood pressure monitor that is a little easier to handle than the cuff type. Not only is it compact in design, but it also comes with its own storage container. The entire device can be kept in this plastic box that is only 3" × 5" × 3".

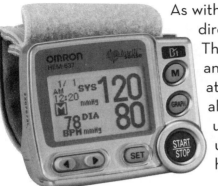

As with any gadget of this type, you must follow the directions carefully to get an accurate reading. The unit fits on your wrist with Velcro grips and is easy to position. The monitor screen is attached to the wrist wrap so that the unit is all in one piece with no cords and is easy to use. The illustrations in the instruction manual show you exactly how to position your hand at heart level. The APS (Advanced Positioning Sensor) starts the monitor only after your hand is in the proper position. This blood pressure monitor has a built-in clock and a 90-memory recall. This allows it to date-time stamp each reading so that you know exactly when each reading was taken.

Model: HEM-637
Manufacturer: Omron Healthcare, Inc.
URL: www.omronhealthcare.com
Price: $99

You can even track your progress on a graph that appears on the monitor screen. The numbers are large and easy to read. Like the LifeSource monitor, the memory function tracks only a single user.

Timex Talking Thermometer

Do you remember having an old mercury thermometer break? You probably snuck in to play with the mercury while your mom cleaned up the broken glass. We have come a long way since those days. Now thermometers are made of unbreakable plastic. They have digital readouts and don't contain dangerous mercury.

This small blue and white thermometer works digitally for accurate measurement. The AccuCurve tip was created to easily find the hotspot under your tongue. The oval shape makes the thermometer easy to use. The thermometer works quickly, giving you a readout in just 30 seconds. You can read the temperature

Manufacturer: Timex
Distributor: Gold Violin
URL: www.goldviolin.com
Price: $19.95

on the small LCD screen, or you can have the temperature read to you in a friendly voice. If you need to read the screen at night, there's even a built-in light.

7-Day Reminder Pillbox Set

It seems that as we age, we take more vitamins and medication. Often it is important to take these pills several times a day. Even the best of us sometimes forget to take our medication. That's why this reminder pillbox set is so important.

It includes seven pillboxes, each with four compartments to hold multiple pills or vitamins. You can put your medication for an entire week in the boxes. Then each morning, you simply slide that day's pillbox into the electronic Reminder. The Reminder has an LCD screen with a large time display. You can set the Reminder to beep up to four different times each day to remind you to take your pills, which are ready and waiting for you to

Model: 7-Day Reminder Pillbox Set
Manufacturer: Gold Violin
URL: www.goldviolin.com
Price: $26.95

remove them from the pillbox. This pillbox is small enough to fit into a pocket or purse.

Independence iBOT Mobility System

They call it a mobility system rather than a wheelchair, but in effect, it is the most high-tech wheelchair ever developed. The iBOT is able to maneuver up and down stairs and to travel over grass, gravel, mud, and sand. It traverses curbs and other uneven terrain easily.

That alone is almost miraculous, but this chair has another outstanding feature. A special balance function allows the four-wheel chair to fold two of the wheels under the other two, lifting the rider up to eye level. This is a real breakthrough for the wheelchair-bound, who are now able to look at the world and have the world look at them eye to eye.

The iBOT is not for everyone. You need to go through a special assessment to see if you are qualified, and then you must be trained on the product. There is a fee for both the assessment

Model: iBOT 3000
Manufacturer: Independence Technology, LLC
URL: www.independencenow.com/ibot
Price: $29,000

and training, but that pales next to the price tag of the iBOT itself. Yet, this is a cutting-edge machine, and it provides the user with experiences that would otherwise be unavailable to him.

Memory and Relaxation Aids

Navigating this fast-paced world often means tension, aggravation, and anxiety. Sometimes the computer and other technological wonders add stress to our lives. So it is only fitting that we also use some high-tech devices to help us relax.

RESPeRATE

More than 50 million Americans—nearly one in four adults—have high blood pressure. This unique FDA-approved device guides the user through simple breathing exercises to slow their breathing. RESPeRATE is based on the fact that breathing exercises can lower blood pressure by relaxing the muscles surrounding the small blood vessels, thereby allowing blood to flow more freely. Amazingly, clinical studies show that this lowering of blood pressure is a sustained reduction. The machine lowers your blood pressure not just when you are in the process of using the machine, but day in and day out, just as a medication would. In fact, RESPeRATE is a gadget, but it acts like a drug. The real advantage of this machine is that it has no harmful side effects.

Model: RESPeRATE
Manufacturer: InterCure
URL: www.resperate.com
Price: $299

The RESPeRATE is pleasant and easy to use. It lowered my blood pressure, but if it doesn't lower yours, you can return it within 8 weeks and pay only the shipping costs.

Sound Soother Fifty

If sound therapy helps you relax, this is the ultimate gadget. It's a bit pricey, but you get a lot for your money. The sturdy unit has aluminum-cone stereo speakers and large buttons that are easy to use. The dog-bone-shaped remote control has a lighted LCD screen and is designed for usefulness and simplicity. The name of the sound scrolls by in the middle of the remote.

With 50 sounds to choose from, there is something for everyone. The sounds include the desert wind, songbirds, crickets, buggy ride, rain forest, brook, steam train, and clothes dryer. I love the brass chimes for listening during the day, and the rainy night sounds can put me to sleep in a flash. I know several mothers who use the white noise to put their babies to sleep. With this system, the babies would probably also love the pink noise and the hair dryer sound.

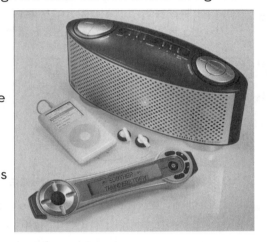

Model: Sound Soother Fifty
Manufacturer: Sharper Image
URL: **www.sharperimage.com**
Price: **$199.95**

The stereo speaker and subwoofer system produces rich, full-bodied sounds. Sharper Image put a lot of thought into this product. It has six presets and six sleep settings. It has an audio-out jack for headphones, and you can hook it up to your home stereo if you like. The Sound Soother even comes with the cable needed to plug in a portable audio device, turning it into a speaker system for your iPod or other digital music player.

Reflexology Foot Massager

It's tough to relax and unwind when your feet hurt. So give your tootsies a treat with the Reflexology Foot Massager. This is a hands-free unit. The buttons are designed to be pressed with your feet. All you need to do is put your toe on a button to have this mechanical masseuse give your feet a treatment. Choose from a gentle or invigorating massage. The unit also has infrared heat, providing soothing heat for worn-out feet.

Model: Reflexology Foot Massager
Distributor: Gold Violin
URL: www.goldviolin.com
Price: $39.95

Journey to the Wild Divine

Journey to the Wild Divine is a software program that comes with its own biofeedback mechanism. The biofeedback regulator attaches to the computer's USB port. Then three small sensors are attached to the fingers to monitor skin conductance and heart rate variability. The Journey is an expedition through a beautiful on-screen world of lush gardens, green forests, and snow-capped mountains.

The trick is that you navigate and move onscreen objects by relaxing or tensing your body. The biofeedback monitors sense your physical state, and the objects on the screen respond to that state. During your journey, you get to juggle balls, move a pair of bellows to start a fire, and levitate objects. You do all this just by controlling your body tenseness and breathing. You never have to touch the mouse or keyboard.

Believe it or not, the program does teach you how to relax, something we seldom do in today's high-speed world.

Model: Journey to the Wild Divine
Manufacturer: Wild Divine Project
URL: www.wilddivine.com
Price: $159.99

Alerts and Emergency Equipment

We all worry about what to do in an emergency. It would be nice if we could purchase an insurance policy that would prevent fires, falls, and other emergencies. Because that is not possible, we have to turn to the next best thing: gadgets that can get us help in emergency situations. Technology comes to our aid with some helpful high-tech wonders.

Telemergency

Telemergency's basis concept is that when connected to your regular telephone line, you can get help by simply pushing a button on the unit or on the wearable pendant. Telemergency automatically dials 911 and up to five additional telephone numbers of your choice for each of three different types of emergencies. A prerecorded emergency message is then delivered.

It is easy to start using the Telemergency. Just plug it into your phone line and into the electrical outlet. When activated, Telemergency emits a sound to let you know help is on the way.

Model: Model 700
Manufacturer: Telemergency LTD
URL: www.telemergency300.com
Price: $229.95

Even better than that, after the called party answers and receives the emergency information message and instructions, they have the option to activate the controlled speakerphone feature. This allows them to verbally communicate with you and let you know help is on the way. You can also use the speakerphone to listen to the area where the base unit is located or to repeat the emergency message. This two-way verbal contact is valuable for sharing critical information and assuring that help is on the way.

Cheaper models without the wireless remote pendant or speakerphone are available. There are no monthly service charges.

Help at Hand Emergency

This is a simplified version of the Telemergency that uses the same concept. Four preprogrammed emergency telephone numbers are automatically dialed and a prerecorded help message is sent when you pull the ring on the wireless pendant. Messages and telephone numbers are customizable. Silent dialing gives added protection during a crime. There are no expensive monthly fees to keep this helper at hand.

Model: Help at Hand Emergency Telephone Dialer
Manufacturer: Maxi-Aids
Distributor: Maxi-Aids
URL: www.maxiaids.com
Price: $99.95

911 Guardian

A cell phone can be a valuable tool in time of emergency. Why not a wearable phone that can put you in touch with help immediately? 911 Guardian is just that. The 3" × 6" device is larger than some cell phones, but it's light and flat enough to be worn around the neck or on a belt. The Guardian operates like a cordless phone with a 2.4GHz operating frequency that lets it cover about 800 feet.

In an emergency, you do not need to be near a landline, and you do not have to be near a base station. You simply press the phone's single button and talk directly to the 911 operator. Because the call goes directly to 911 rather than to a call center, there are no monthly charges. The phone's lithium battery lasts for about a year, so the phone does not need to be recharged.

Model: 911 Guardian Phone
Manufacturer: DesignTech International, LLC
URL: www.designtech-intl.com
Price: $139.95

Philips HeartStart Home Defibrillator

Approximately 340,000 people in the United States die from sudden cardiac arrest (SCA) each year. It can happen any time, any place, and nearly 80% of all sudden cardiac arrests happen at home. A defibrillator is a device used to treat the most common underlying cause of sudden cardiac arrest. For the best chance of survival, a shock from a defibrillator should be delivered within 5 minutes. For every minute that goes by without treatment, a victim's chance of surviving SCA decreases by 7–10%. After 10 minutes, survival is highly unlikely.

A few years ago, these devices were available only to healthcare professionals and emergency personnel. Now, however, the FDA has cleared the Philips HeartStart Defibrillator for home use without a prescription. HeartStart was designed so that virtually anyone can use it to help save the life of a person who suffers a sudden cardiac arrest. It also has voice instructions that talk you through each step and coach you through the CPR. The device uses technology and accurate algorithms to determine whether the patient's heart requires a shock from the defibrillator. Designed for safety, HeartStart analyzes a patient's heart rhythm and will recommend a shock only if necessary. There's no doubt about it: As more homes acquire the HeartStart Home Defibrillator, more lives will be saved.

Model: Philips HeartStart Home Defibrillator
Manufacturer: Philips
URL: **www.heartstarthome.com**
Price: $1,495

MedKey

The MedKey is a USB hard drive that is used to store patient medical information, including emergency contact information, diabetic conditions, allergic reactions, prescription dug information, and medical records.

Here's how it works. You purchase the MedKey Package and pay a small annual fee for backup storage. Then you enter and approve all the health information, which is then encrypted on your MedKey device. A copy is kept at the encrypted MedBase.com patient data repository, which is accessible by approved medical personnel. You then carry your MedKey device on your key chain or in your pocket. In case of emergency, authorized personnel can plug your MedKey into their computer, PDA, or smart phone to access your records and be able to provide appropriate emergency treatment. You can update your MedKey records by plugging the MedKey into your computer.

Model: MedKey
Manufacturer: MedKey Corporation
URL: www.medkey.com
Price: $69.96

Although MedKey is in its infancy, it is a good stab at portable storage of heath records. It is poised to be the wave of the future.

Sandy's Lingo List

The world of technology has created some crazy new words. Here are explanations for a few of the more unusual words used in this chapter.

body mass index (BMI)—A measurement of the relative percentages of fat and muscle mass in the body. It is often used as an index of obesity.

hypertension—The medical term for what is commonly known as high blood pressure.

pedometers—Small devices that measure how far a person walks or runs.

ultraviolet (UV) rays—The part of the sun's spectrum that are invisible to the human eye. UV sunlight in excessive amounts has been shown to be harmful to human skin.

Chapter 7

Communication Devices

Think like a wise man but communicate in the language of the people.

—William Butler Yeats

Technology and communications go hand in hand. This dynamic duo has made a great impact in our lives over the years. The telegraph led to the telephone, which in turn led to the Internet. Along the way, our lives have been improved with television, fax machines, and cell phones. Although some aspects of technology draw people away from each other, communications help people stay in touch and in tune with each other.

Cell Phones and Accessories

The cell phone might be the gadget that has changed our lives the most in the past few years. Cell phones have become an integral part of most of our daily lives. They give us a newfound feeling of personal power and have also made a profound impact on our social communication. They give us a new means of expressing ourselves. And we can do all of this while we are on the go.

Motorola Cell Phone

The problem with most cell phones today is that everything is too small. Yes, it's nice to have a lightweight phone that fits in your pocket, but how about larger keys and a screen you can see without squinting? I haven't yet found the perfect cell phone, but this Motorola model has decent keys and two readable screens. The main screen is about 1 3/4" × 1 1/2". The display on the front of the phone is larger than most at 1 1/4" by 1".

I also like the 1.2-megapixel camera, the integrated MP3 player, and the speaker phone capability. This phone is the first Verizon phone to have Bluetooth connectivity. With a Bluetooth-enabled phone, you can use handy wireless headphones and hands-free devices.

If you are really into cell phones, you might also appreciate the polyphonic sound that allows for wonderful ring tones and the picture caller ID on both external and internal displays that lets you

> **Model:** V710
> **Manufacturer:** Motorola/Verizon
> **URL: www.motorola.com**
> **Price:** $389

see a picture of the caller. It also has advanced camera features like video clip capture, adjustable resolution, and day/night lighting modes.

Keep in mind that cell phone manufacturers make different models for different cellular providers. The V710 is made to go with the Verizon service. If you have Cingular, T-Mobile, or another service, check its lineup for a similar model.

RazrWire™ Bluetooth Eyewear

Reproduced with permission from Motorola, Inc. © 2005, Motorola, Inc.

Get Smart fans your time has come. Today's electronics are being built into everyday items that will make Maxwell Smart's shoe phone look like old hat. Now you can talk on the phone through your sunglasses with Oakley sunglasses equipped with Motorola Bluetooth wireless technology. A small control module is mounted on the sunglass stem. It includes an earpiece and microphone, two volume buttons, and a multifunction button to handle incoming and outgoing calls. Rechargeable batteries are said to last for five hours of talk time. They can be charged with the included wall charger or through a USB cable linked to a computer. The RazrWire will be available with Cingular wireless service.

The rimless sunglass frame is made of Oakley's proprietary O-Luminum, an alloy 40% lighter than titanium. Blue light and 100% of the harmful UV rays are filtered out by Oakley's Plutonite lenses. The frames can easily be fitted with prescription lenses.

Model: RazrWire™
Manufacturer: Motorola/Oakley
URL: www.oakley.com
Price: $295

A Bluetooth-enabled cell phone hooks up wirelessly to the controller on the sunglasses. The cell phone can be kept in your pocket or anyplace up to 30 feet from the sunglasses. The phone can be set up to ring in your ear so that no one else hears the ring. You can answer the phone by pressing the multifunction button on the sunglasses. If your cell phone supports voice dialing you can dial by simply speaking the name of the person you want to call. What fun. Wearable technology is here!

Bluetooth Easy Install Car Kit

Holding a cell phone up to your ear when you're in the car forces you to take one hand off the wheel. With this car kit, you can talk hands-free. Just charge up this small Bluetooth device, and then clip it to your visor. The speaker includes duplex, echo, and noise cancellation technology, so voice quality is good. You can talk for about three hours on one charge.

This streamlined device has only two buttons. The multipurpose button on the top answers, ends, and initiates calls. It also controls call-holding and 3-way calling. The button on the side turns the speaker on and off and controls the volume.

You have to set up your Bluetooth-enabled Motorola cell phone to communicate with the speakerphone. You also have to initiate the Bluetooth connection when you want to use the speakerphone. This can be set up to be a single button pressed on the phone. Although the steps to making this happen are not terribly complicated, you do need to be familiar with your phone's settings. If you have a problem, the store where you purchased your cell phone or the Bluetooth kit can help you.

Model: HF 800
Manufacturer: Motorola
URL: www.motorola.com
Price: $119.99

Cellboost

Sandy's favorite

All wireless cell phones have one thing in common: They are powered by rechargeable batteries, and it is up to the user to keep those batteries charged. As you know, when dealing with computers, cell phones, and other high-tech equipment, Murphy's Law always applies. If anything can go wrong, it will. So there will, no doubt, be a time when you need to make an important phone call and find that your cell phone batteries are too low.

That's when you need the Cellboost, a disposable battery with a special plug that fits into your cell phone. The Cellboost provides approximately 60 minutes of talk time and 60 hours of standby time in a tiny, lightweight hard plastic container. At 2 1/2" × 1 1/4" and only 1/4" thick, you can keep the Cellboost in your briefcase, pocket book, or vehicle's glove box for emergency power. The Cellboost dispenses its energy through a "quick-charge" method that transfers full power to your phone quickly and easily.

You need to purchase a Cellboost that is compatible with your cell phone manufacturer. Each cell phone type has a different colored cover, so after you have determined the type you need, it is easy to pick up another one. Cellboost is available at drugstores, supermarkets, and airports.

Model: Purchase model based on type of cell phone
Manufacturer: The Compact Power Systems, Inc.
URL: www.cellboost.com
Price: $5.99

TracFone

In the past several years, prepaid telephone cards have become popular. Take that concept one step further, and you have the TracFone, a cellular phone that lets you pay as you go. TracFone eliminates the complex service contract, which often binds you to long-term payments. There are no monthly bills and no credit checks on this pay-as-you-go system.

Here's how it works. You purchase a TracFone mobile telephone. You have a choice of several different manufacturers, such as Motorola and Nokia. Various models are available at a variety of prices. You get the phone, battery, and charger. You also get 60 days of

Model: Motorola c155
Manufacturer: TracFone Wireless, Inc.
URL: www.tracfone.com
Price: $59.99

wireless service (starts at activation) and 10 minutes of starter airtime. Then you purchase wireless prepaid airtime cards. These are sold in increments of 40, 100, 200, and 400 units for approximately $19.99, $29.99, $49.99, and $79.99, respectively.

One nice feature is that the number of minutes you have remaining appears right on the screen of your mobile phone. When you are running low, you simply purchase another card. It is easy because the cards are sold in many locations, including Eckerd Drug, Kmart, Lowes, RadioShack, Office Depot, Safeway, Staples, Target, Wal-Mart, Western Union, and 7-Eleven.

Plantronics Voyager 510 Headset

Bluetooth technology allows you to talk on your Bluetooth-enabled cell phone using wireless headphones like the Plantronic Voyager 510. This comfortable headset hooks over one ear and has a small microphone attached. The noise-canceling microphone offers great audio quality. If you communicate by phone a lot, you can use the deskphone adapter to turn your home or office phone into a Bluetooth phone. Then you can switch between your cell phone and your desktop phone using the headset. It's pretty cool technology. It comes with a recharging base. The headset folds to fit your pocket, so you can use it anywhere. With up to 6 hours of talk time, you can talk to your heart's content.

Model: Voyager 510
Manufacturer: Plantronics
URL: www.plantronics.com
Price: $99.99

CellDock

CellDock is a cellular accessory that allows you to integrate your cell phone with your landline phone so that you can make and receive calls from your cell phone by using a regular telephone. You can also use CellDock to connect corded telephones to your cell phone if you have joined the growing trend of completely eliminating your landline telephone service.

This device is a docking station for your cell phone. You simply hook the CellDock to your telephone or your telephone landline. Then insert your cell phone into the dock. An incoming call to your cell phone rings on the cell *and* the attached telephone. To place an outgoing call, press the pound sign before and after the number to use your cellular service, or just dial to place the call from the regular telephone line.

One caveat is that you must connect the CellDock to the main telephone line coming into your house to have incoming cell phone calls ring on all your phones. If you don't know which is the main line, you might have to move the CellDock from outlet to outlet to find it. The CellDock comes with all the adapters you need for different brands of cell phones. One nice additional feature is that the CellDock is also a charger for your cell phone.

Model: CellDock 1000
Manufacturer: CellDock
URL: www.celldock.com
Price: $99.99

CellAntenna

Cell phones are great for everyday communications and for emergencies. Unfortunately, in many areas of the country, cell phone reception is spotty. If you live in one of those areas or you travel a lot or have an RV, you can give your cell phone reception a boost with an antenna.

The CellAntenna is nice because the 3" magnetic base with rubber shield allows you to easily attach it to your car without damaging the car's finish. You can place it on the trunk, or for better reception, on

Model: CA4MWM
Manufacturer: CellAntenna Corporation
URL: www.cellantenna.com
Price: $49.95

the roof. Then you route the cable through a slightly open window or the molding of a door or hatchback.

An optional $12.95 adapter that is cell phone-specific is then attached to the cable and to the cell phone. You might have to refer to your phone documentation to find the antenna jack on your phone. Other than this, installation is easy.

This kit includes 2", 4", 12", and 24" whip antennas that are interchange-able. You can experiment to see which one works best in your location, or you can use the shorter one for city driving and the longer one when get out in the country. It is nice to have all four antennas. I know I am being picky, but because it's possible to use only one antenna at a time, I was left wanting a little case in which to store the extra antennas.

Telephones and Enhancements

Many of us remember rotary dial telephones and exchanges like Pennsylvania-6, Independence-3, and Brunswick-8. Touch phones and area codes are only a few of the technologies that have affected telephone communications. Today we have wireless phones and caller ID. We also have telephones with special devices to help those with vision and hearing impairments. All of these new telephones and devices help to make our everyday lives easier.

Bright ID'er

If you love using the caller ID function on your telephone to see who is calling, but you can't deal with those tiny letters and numbers on the telephone display, this product was made for you. The Bright ID'er is the largest caller ID display that I've seen. The screen is about 8" long and more than 1" high. The name or telephone number of the caller appears on the screen in bright red LED numbers and letters that are easy to see in a darkened room or from across a room. Bright ID'er is AC powered and can store the

Model: Bright ID'er
Manufacturer: Independent Technologies
URL: **www.brightider.com**
Price: $109.95

name, number, and time of 99 calls. The entire memory can be erased, or you can erase individual calls as you like. One unique feature is this device's ability to indicate when an extension is in use or left off the hook or if basic service has been interrupted.

You need to have caller ID service from your telephone company to use Bright ID'er. The Bright ID'er is good looking enough to be placed in any location that offers the best viewing. You can even mount it on the wall. With this display, you will always know who's calling.

Olympia InfoGlobe Plus

The InfoGlobe Plus is a distinctive device. It combines a cordless telephone with a 54-memory caller ID and message display. This gadget's uniqueness, however, is in the lighted blue globe that functions as a display. Like a digital news ticker, the name and telephone number of the caller scrolls around the globe. You can also program the InfoGlobe to display your own message. In fact, you can create six different messages.

The display is an attention grabber and is somewhat mesmerizing. Although it is fun to look at and to program with your own messages, InfoGlobe is also a serious caller ID unit and can be used just for that. You can see who is calling or look at who has called and erase messages just like any other caller ID device. The large blue letters make it easy to read the InfoGlobe even from across the room.

This is one smart gadget. It has a 100-year internal calendar with auto adjustment for Daylight Savings. It exhibits a holiday greeting on New Year's Day, Father's Day, Mother's Day, and other special days. And it has foreign language characters, so it can display messages in several different languages.

Model: OL 3020
Manufacturer: Wave Industries
URL: www.olympiaphones.com
Price: $99.95

The wireless telephone is an excellent 5.8GHz 40-channel phone with a nice feel and voice quality. However, it has fairly small buttons, and the screen on the phone is difficult to read. If you don't mind that, this is a good package. If you are just looking for the wonderful globe-like Caller ID, you can purchase the InfoGlobe alone with similar features for about half the price.

Amplified Big-Button Talking Phone

The big buttons on this phone are so easy to use that you'll wonder why every telephone doesn't have large buttons. The buttons are also labeled with Braille characters. As you dial, the "talking" buttons can repeat each number.

This phone has adjustable amplification. You can make calls up to 70 times louder, or you can adjust the volume to the degree of loudness that is comfortable to you. This Big-Button Phone also minimizes background noise and enhances speech clarity for those who suffer from hearing loss. It can boost the volume of outgoing speech, too.

The phone has a flasher to let you know when it is ringing. The handset is hearing aid T-coil compatible. All the buttons and volume sliders are easy to find and use. In addition to these special features, this phone also has all the regular telephone features, including 10 programmable memory buttons and 3 programmable emergency buttons.

Model: Amplified Big Button Talking Phone
Manufacturer: Gold Violin
URL: www.goldviolin.com
Price: $149.95

Loud & Clear Cordless Phone

Sandy's favorite

This 900MHz cordless phone has large buttons and a clear screen. The bright red adjustable visual ringer lets you see *and* hear when a call is coming in. An audio boost button is also available. The phone offers separate ringer and earpiece volume controls, one-touch emergency dialing, and a 10-number memory.

Besides the nice lighted screen and keys, this phone has a few other useful features. As long as you subscribe to call waiting and caller ID, when you are on a call and another call comes in, the caller's name and number scroll across the screen. The phone also has a key to press on the base to help you locate an errant handset. The Uniden UltraClear Plus technology eliminates background noise. With 20 different channels to choose from, you can always find a clear channel.

> **Model:** EZI 996
> **Manufacturer:** Uniden
> **URL: www.uniden.com**
> **Price: $39.99**

Reizen Speak N' Hear Voice Amplification System

This system is designed to amplify the voice of those who have temporary or permanent voice impairments. It is lightweight and portable, with good quality performance. You can even use it for projecting your voice outdoors or for speaking to a crowd.

The system comes with a nylon adjustable waistband so that you can wear it around your waist. It also has a lapel microphone and a headset microphone. You can use either with the unit.

> **Model:** Reizen Speak N' Hear Voice Amplification System
> **Model:** RE-120
> **Manufacturer:** Reizen
> **URL: www.maxiaides.com**
> **Price: $159.95**

Vocally Voice-Activated Phone Dialer

This neat little gadget lets you speak a person's name and automatically dials that person's preprogrammed telephone number without your having to fumble with the number key pad on the telephone.

Connect this small rectangular box to the telephone line and to the telephone, plug it into the wall, and you are ready to go. The Phone Dialer's voice guides you in setting up its interactive voice menu and helps you add, delete, or hear a name. Although you can get through the setup without a manual or quick start guide, I would like to see one included.

After the Phone Dialer is set up, you can simply speak a name, and your phone dials that person's number. It remembers up to 60 names and numbers. The Dialer can handle telephone numbers up to 19 digits long, so you can add overseas numbers with ease. It works with any telephone and recognizes any language. Overall, it's a pretty smart device!

> **Model:** Voice-Activated Phone Dialer
> **Manufacturer:** Maxi-Aids
> **URL:** www.maxiaids.com
> **Price:** $199.95

Specialty Communication Devices

Cell phones and wireless phones have become a mainstay in our lives. Yet many other types of communication devices are useful. If you want to communicate with someone nearby, a simple set of walkie-talkies might suffice. When you really want to see that new baby or college student in another state, look to more high-tech devices like Web cams or videophones. Technology has something for everyone.

Beamer TV Videophone

This device turns your telephone and television into a videophone. You set the small, sturdy Beamer TV box on top of the television. You connect it to your regular telephone line and to any regular-corded or cordless telephone with the included telephone cables. Then you run a video cable from the Beamer to the RCA jack that is standard on most televisions. Setup and use is straightforward. After you've set up the device, you simply dial the number. When the other party answers (that person must also have a Beamer), you press the button on the Beamer remote control to connect the video. In about a minute, you will see the other party on your television screen. You will also see yourself in a smaller square on the TV screen.

Model: Beamer TV Videophone
Manufacturer: Vialta
URL: **www.vialta.com/beamertv.htm**
Price: $124.99

Because the video transmission is accomplished over regular analog telephone lines, the picture is a little grainy and somewhat choppy, but you really can see each other. If you do decide to try a Beamer, you should know that the lighting makes a big difference. In a bright environment, you get a much better picture.

You might be willing to put up with the slightly choppy video when you realize that there is no special wiring, no additional telephone charges, and no monthly fees. On a recent call, when my daughter commented on my new hairstyle, I realized that we really were seeing each other over the telephone. Wouldn't Alexander Graham Bell be amazed?

Vialta also makes a video phone with a screen that can be used just like a telephone.

iZon Hands-Free Walkie Talkie

In many circumstances, a pair of Walkie Talkies can make communicating easier. You can use them at home from room to room, on the road when traveling with more than one car, or in the mall from person to person.

I've found some Walkie Talkies to be difficult to use. This set of IZon Walkie Talkies is different. You get two units with large buttons that are easy to use. The black devices are on a strap and worn just like a wristwatch. In fact, the default setting shows the time, so they actually are wristwatches with added features.

Model: 05-1001
Manufacturer: Global American Technologies
URL: http://www.globalat.com/ walkietalkie.html
Price: $99

Although a bit chunky, they are lightweight and look pretty good. Each unit includes an LCD screen, built-in speaker, and built-in microphone. A silver button on the top allows you to talk to the other person. The cool thing is that you can also talk hands-free. When you're wearing the watch, you simply lift up your wrist until it is against the watch and talk.

These Hands-Free Walkie Talkies work on rechargeable lithium-ion batteries. Two chargers are included, so you can charge both units at the same time. They have a range up to 1.5 miles and work on 22 different channels.

Live Ultra Web Cam

If you are far away from loved ones, a Web cam lets you see them as you chat over your Internet connection. It's perfect for video instant messaging. This particular Web cam has some great features. The camera itself uses a CCD (charge-coupled device) sensor that produces better resolution and color.

Sandy's favorite

If you have a large family or group of people who want to get in the picture, the Live Ultra has a wide-angle lens that captures a 50% wider field of view than most Web cams. The Smart Face tracking is pretty cool. It tracks your movements and keeps the camera focused on your face. The included headset is small and unobtrusive, and it produces much better sound than built-in microphones.

If you have the faster USB 2.0 connection that is on most newer computers, the video is smoother. However, you can still use this Web cam even if you have an older computer with USB 1.1. The included software is impressive. It helps you make mini-movies, send digital post-cards featuring your own videos, and even lets you do time-lapse video capture. The Live Ultra also acts as a motion-activated security device.

Model: Live Ultra Web Cam
Manufacturer: Creative
URL: www.creative.com
Price: $99.99

i2eye Videophone

Would you like to turn your TV into a broadband videophone? Don't have a computer? No problem with the D-Link i2eye. All you need is a broadband Internet connection, a TV, and the i2eye.

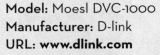

This small device makes it easy to keep in touch with family and friends around the world. The i2eye has a built-in microphone and an adjustable lens. The quality of the video depends on the quality of your broadband connection. Because broadband has a bigger pipeline than a regular telephone connec-

Model: Moesl DVC-1000
Manufacturer: D-link
URL: www.dlink.com
Price: $299 each, $499 a pair

tion, the picture quality far surpasses videophones like the Beamer, which use slower telephone connections. In fact, the video is quite good.

The thing I like about this videophone is the ease of setup. A Setup Wizard walks you through, with the setup being painless compared to other videophones. The unit comes with a remote control so that you can switch from the Picture-in-Picture mode to a Full Screen display quickly and easily. You can even personalize your display.

Another great feature is that D-Link has a server that registers a "phone number" for your videophone. When anyone who has an i2eye calls that number, your i2eye is "called." There's no playing around with changing IP addresses that many other videophones use.

Sandy's Lingo List

The world of technology has created some crazy new words. Here are explanations for a few of the more unusual words used in this chapter.

Bluetooth—A short-range technology that uses radio frequencies to connect devices wirelessly. Bluetooth technology is embedded in a computer chip and can be used in many different devices. A Bluetooth camera can transmit its pictures to a Bluetooth computer or printer without being attached.

charged coupled device (CCD)—A sensor that records images. It is a component of many digital cameras.

web cam—Short for *web camera*, a web cam is a camera that attaches to the computer. With a camera and software on each computer, two or more people can communicate visually and audibly over the Internet.

Chapter 8

Gadgets for the Home

If a man can write a better book, or preach a better sermon, or build a better mousetrap than his neighbor, though he builds his house in the woods, the world will make a beaten path to his door.

—Ralph Waldo Emerson

You can find plenty of gadgets and gizmos to entertain you and help you communicate with others, but what's available for you when you're at home? After all, home is the place that you spend the most time. It is often the place where a little help is most appreciated. We don't yet have robotic butlers or housekeepers, but we are moving in that direction. In the meantime, we have plenty of gadgets that can eliminate some of the small frustrations and add to life's enjoyment.

Organize Your Life

Book after book has been written about how to organize your life. All this help is offered because it is difficult for the average person to stay organized. But help is at hand with a few high-tech gadgets that can aide you in managing all your stuff.

Now You Can Find It

Why am I always losing things? Is it stupidity? Absentmindedness? The beginning of Alzheimer's disease? Whatever the cause, technology has addressed my problem with the Now You Can Find It gadget.

Now You Can Find It consists of a base station with eight differently colored buttons and eight color-coded key-chain style disks. A magnetic mounting bracket that allows the base to be kept on the refrigerator or other metal cabinet is also included. Attach the orange disk to your car keys. Then when you misplace your keys, press the orange button on the base. The orange disk on the keys responds by flashing and beeping. This device works by radio frequencies, and the signal is said to reach 30 feet. In reality, you probably need to walk around the house pressing the button several times before you find your keys. But it sure beats frantically looking for 15 minutes while you get later and later for your appointment.

Model: Ultra 8
Manufacturer: Sharper Image
URL: www.sharperimage.com
Price: $69.95

The size of the disks and key-chain–like method of attachment limit the types of objects that can be used with this device. It is perfect for my keys and USB hard drive. I keep one on my purse and one on my gym bag. It worked great for my PDA, which has a loop on the case. It was not, however, useful for my eyeglasses or the television remote, which are the other two elusive items that are often difficult to find.

LetraTag

Nothing can make you feel more organ- ized than having everything categorized with nice neat labels. The LetraTag is just what you need to make that happen. Although not terribly high-tech, this little labeler sure beats the old-fashioned labelers where you have to turn the disk to locate the letter and press the handle to imprint the raised letter on the tape.

Sandy's favorite

The LetraTag is a standalone tool. It is highly portable and fits nicely in the palm of your hand. It is easy to find the letter you need on the ABC-ordered keypad. Just type in what you want on the label, and press the Print button. When the tape emerges from the machine, press the Cut button, and the label is done. The tape backing is per- forated, so there is no hassle in removing the backing when you are ready to stick on the label.

Model: LetraTag
Manufacturer: Dymo
URL: www.dymo.com
Price: $44.99

This labeler handles tape up to 1/2-inch wide. The tapes come in a variety of colors and types including paper, plastic, metallic, and iron-on fabric tape. This little device is so useful that I have gone hog-label-wild. I now have labels on my files, pantry food containers, storage boxes, and bookshelves. Every day, I find more things to label. It makes me feel so organized!

New Dymo Labeler

If you have ever used a label maker that attaches to the computer, you know they are useful. When you make an address label, you can store the name and address for future use. You can also pull a name and address from your address book program. Making file folder labels is

easy. Instead of printing one at a time like you would with a handheld labeler, you can print as many as you like in succession.

The major problem has always been that when you want to change the size of the labels, say from address labels to file folder labels, it was a messy job that often resulted in ruining several perfectly good labels. Dymo has solved this problem with the new model. It can hold two different label sizes at the same time.

Model: LabelWriter Twin Turbo
Manufacturer: Dymo
URL: www.dymo.com
Price: $179.99

The Twin Turbo uses the same engine as the LabelWriter 400 Turbo, which is the fastest personal label maker on the planet. Because the Twin holds two rolls of tape at one time, it is like having two LabelWriter 400 Turbos on your desktop in a smaller package that costs less. The Twin Turbo produces clear text on a single label in just one second. Labels are automatically aligned. If you make a lot of labels for home or office, there is a lot to like with the Twin Turbo.

SCOTTEVEST

High-tech gadgets are great, except when you have too many to carry. The SCOTTEVEST solves this problem with its technology-enabled clothing. There are vests, jackets, pants, and even hats in this lineup. All have hidden pockets. The Version Three.0 Cotton jacket has 32 pockets—enough to make even the most enthusiastic gadget guru smile. The pockets vary in size to accommodate objects such as ID, cell phone, PDA, business cards, file folders, ear buds, and laptop computers. Each pocket has a Velcro or zipper closure.

Model: Version Three.0 Cotton
Manufacturer: SCOTTEVEST
URL: www.scottevest.com
Price: $169.99

For the high-tech gurus, the jacket includes a hidden Personal Area Network (PAN), which permits wired devices to run through the lining of the jacket from pocket to pocket and to the collar loops that can hold ear pieces or headphones. This allows the wearer to use an electronic gadget like a phone, PDA, or music player while having the device secured in the pocket of the jacket. Real gadget mavens will want to try the optional solar panels that fit on some of the jackets. Just snap them to the back of the jacket, plug in your equipment, and you can let the sun recharge your gadgets on the go.

For those who find themselves carrying a large variety of gadgets, the SCOTTEVEST can easily be a part of their everyday apparel. Although these jackets are made to keep gadgets readily available, the non-tech person will find them useful to carry things like water, medication, and magazines. I love mine for air travel. Not only can you have everything easily accessible, but when you get to security, you can just take the jacket off and put it on the conveyer belt. Amazingly, even with pockets full, the jacket hangs nicely and doesn't look nerdy.

By the way, most of the jackets have removable sleeves so that you can easily turn the jacket into a vest, which is what gave the company the name SCOTTEVEST.

NeatReceipts

Every time you make a purchase, you get little paper receipts. If you are anything like me, all those receipts just accumulate on a table or in a drawer. I know I should keep track of how much I spend. I should keep track of credit card receipts and hold receipts for tax purposes. But somehow I never get around to it. Technology to the rescue! The NeatReceipts package includes a

Model: NeatReceipts
Manufacturer: NeatReceipts
URL: www.neatreceipts.com
Price: $229.95

small color sheet-fed scanner and the software you need to organize your receipts.

When you come home with a receipt, you simply feed it into the scanner. The scanner auto-sizes and auto-crops the receipt and then transfers the information to the computer. The software is smart enough to read and understand the important information from the receipts. It keeps a copy of the original receipt. Then it takes the information and organizes it into a searchable expense database. It even lets you add comments and tax information.

The program also integrates with Quicken, QuickBooks, Money, Excel, and many other programs. It can help you create an expense report. It even allows you to scan and organize other documents like articles, contracts, and faxes. What an easy way to get organized!

Olympus Voice Recorder

Whether you need to keep track of everything covered in that business meeting, you finally have time to catch some lectures, or you just want

an easy way to document your thoughts, this voice recorder will do the trick. At less than 3 ounces, it's lightweight. It's also small enough to fit in a shirt pocket.

Not your father's voice recorder, this new digital recorder is a capable, high-performance instrument. It has several recording modes and data compression formats that allow for up to 22 hours of recording time with up to 995 individual recordings. You can use some cell phones and PDAs for recording quick notes, but this dedicated recorder is worlds above those devices. It captures clear voices no matter what the venue. The recordings in a concert hall are just as clear as those of a personal conversation. The DS2 also lets you set index points when you know you will want to go back to a certain part of the recording. The USB docking station makes it easy to upload your recordings to the computer.

Model: DS2
Manufacturer: Olympus
URL: **www.olympusamerica.com**
Price: $149

You can also edit them with the included software. The recorder can be voice-activated, and it records in stereo.

Basic navigation like record, fast forward, and so on is handled by good-sized clearly marked buttons on the front of the recorder. You will be able to handle basic functionality without a manual, but I advise looking at the manual to take advantage of all of this product's functionality.

iTalk iPod Voice Recorder

If you own an iPod and want to do some voice recording, this simple iPod add-on is just what you need. At 2 1/2" × 3/4" × 1/2", this miniscule device won't add much heft to your iPod. It simply plugs into the top of the iPod and turns it into a voice recorder. The built-in microphone does a good job recording. If you like, you can also connect an external microphone directly to the iTalk.

The iTalk is so small that it is hard to imagine that the mini-speaker would sound so good. It does a good job replaying recordings, and you can use it for listening to audio books. In a pinch, it can function as a speaker for playing the music on an iPod.

Model: iTalk
Manufacturer: Griffin Technology
URL: **www.griffintechnology.com**
Price: $39.99

You can listen to your voice recordings through your headphones. There is also a pass-through mini jack that allows you to monitor your voice recording or listen to music without removing the iTalk. In addition, you can download your voice recordings to your computer.

This gadget is great for recording a grocery list or documenting that great idea before you forget it. With iTalk, you can use your iPod to record everything from a lecture to a conversation.

High-Tech Helpers

The gadgets that I really love are those that alleviate some of life's little frustrations or that actually help me accomplish a task around the house. Here are a few of my favorites.

Magic Tap

It never fails. When I enter the bedroom at night and switch on the lamp from the wall switch, it doesn't go on because it was last turned off at the lamp. Or, when I try the switch on the lamp, it doesn't go on because it was last turned off at the wall switch. Magic Tap solves both of those problems, plus it

Sandy's favorite

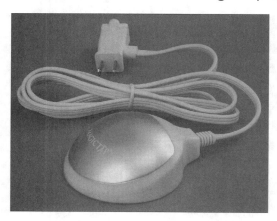

Model: Magic Tap Basic
Manufacturer: Imagine-It-Concepts, Inc.
URL: www.imagine-it-concepts.com
Price: $24.95

makes it easier to control hard-to-reach or difficult-to-operate lamp switches. Magic Tap turns any lamp into a touch-controlled light, which is always ready to be turned on with a tap. The included 3-way switch adapter returns the wall switch to the on position after every use, creating a situation where the lamp can always be turned on from both the wall switch *and* at the lamp, no matter where it was switched off. Because you can control the light with a simple tap, you never have to fumble with the switch on the lamp itself.

The Magic Tap comes in two versions: Basic and Deluxe. They work similarly, and both include the wall switch adapter. Magic Tap Basic looks like a small, round half ball that has been slightly flattened. The sides are slightly curved. There is a small blue light near the cable that makes it easy to find in the dark. Happily, the light is not bright enough to be bothersome for light sleepers.

Magic Tap Deluxe is a small translucent picture frame with blue backing that gives off a soft blue glow in the dark. The triangular shape of the back of the frame allows it to rest sturdily on the table or nightstand. You can remove the translucent frame and easily insert any picture you like.

Electric-Powered Chair Riser

Sometimes age, pain, or weakness make it difficult to get in and out of chairs. Don't worry. There's help available with the Electric-Powered Chair Riser. Plug it in, and place it on any sofa, armchair, or recliner. As you sit down, the Chair Riser senses your weight and gently lowers you into place. When you are ready to get up, the Chair Riser helps lift you to the standing position. The navy blue cover is machine washable, and the cushion itself is made of high-density foam. This is just the ticket when you need a little boost.

Model: Electric-Powered Chair Riser
Distributor: Gold Violin
URL: www.goldviolin.com
Price: $189

LavNav Lavatory Navigation Night Light

Ever since bathrooms moved indoors, men and women have been at odds over the position of the toilet seat. Just about every woman in America has at one time or another taken a sleepy trip to the bathroom only to "fall into" a toilet when the seat has been left up—an uncomfortable and potentially injurious scenario. Men feel badgered by the constant "Honey, would you please put the toilet seat down?" mantra from the women in their lives.

LavNav to the rescue! This small, contoured unit attaches easily to the toilet lid with the enclosed adhesive

Model: LavNav Lavatory Navigation Night Light
Manufacturer: Arkon
URL: www.arkon.com/ toiletnightlight.html
Price: $29.95

strip. Its built-in sensor is triggered by movement and assess the position of the seat. When you walk up to the toilet in a darkened room, it lights up green if the toilet seat is down, or red if it is up. You might be laughing, but this is a serious product. It also provides smart, energy-efficient lighting for those sleepy late-night excursions to the bathroom. No need to turn on that startling bright light. Another plus is that LavNav is powered by two AA batteries, so it doesn't take up any of the bathroom electrical sockets that always seem to be in short supply.

Fellowes Paper Shredder

Sometimes identity theft occurs from online sources, but often the information that is needed to perpetrate this devilish act comes from offline sources like stolen wallets and garbage cans. Shredding personal documents before you throw them in the trash is a good way to protect yourself from identity theft.

This Fellowes shredder is a strong shredder that is perfect for home use. It has a 9-inch opening that accommodates most documents, and it can handle up to 10 sheets at a time. Instead of shredding into strips, it gives the added security of a cross-cut, which shreds the paper into small pieces.

Model: Powershred P-55C
Manufacturer: Fellowes
URL: www.fellowes.com
Price: $79.99

This unit has a safety lock for the times when little ones might be around. It also has an opening on the front of the five-gallon wastebasket that can be conveniently used for nonshredded trash.

Data Destroyer

With more and more information being stored in computers, security is constantly an issue. We tend to think about computer security more than we do about the security of our data when it is not on the computer. Today, most people use CDs to back up their computer data. As a precaution, they might even make duplicate copies. If these data discs contain financial and other personal information, you don't just want to throw them in the garbage where anyone can pick them up.

Model: DD3001
Manufacturer: Norazza
URL: **www.norazza.com**
Price: **$49.99**

To protect yourself, you need to completely obliterate the data. That's where the Data Destroyer comes in handy. This little gadget is like a paper shredder for CDs and DVDs. And like a paper shredder, it is easy to use. Just insert a disc into the slot, and the machine does all the work. The Destroyer doesn't shred the disc, but it completely damages both its sides. When the disc emerges from the Destroyer, you see what looks like small dots all over the disk. This is an indication that the disk is completely unreadable with all data destroyed.

If you are in a hurry or have a lot of discs to destroy, this unit can handle 15 discs per minute. Whether you use it for 1 disc or 100, the Data Destroyer can help you protect your privacy and keep identity theft at bay.

Home Heartbeat

Rather than a home security system, this is a home awareness system. It monitors items of your choice. Want to make sure the iron is turned off? Just plug the iron into a Home Heartbeat sensor before you plug it into the wall. Then the status of the device appears on a small key chain-type device, called a Home Key, that you take with you.

The Home Heartbeat comes with a base station with one sensor for $149. Additional sensors range from $29 to $49. Sensors are

Model: Basic Home Heartbeat Kit
Manufacturer: Eaton
URL: www.homeheartbeat.com
Price: $149.99

available for many different tasks. One tells you if a window, door, or garage door is open or closed. Another tells you if it senses water. You can even purchase a water valve shutoff to shut off the water in case of potential flooding. The water valve shutoff might require installation by a plumber. All others are easy to install.

This system is a flexible alert system. An attention sensor lets you send messages to the Home Key device. A reminder sensor helps you remember important tasks. If you don't want to take the time to check the screen on the Home Key for the status of the sensors, an additional alert system is available. For a small yearly fee, if a sensor is triggered, it can send you a notification by email or cell phone.

Eaton is dedicated to expanding and improving this system. Proposed future enhancements include motion, gas detection, and temperature sensors.

Domestic Appliances

When electricity started to be used in homes, people were able to stop beating rugs and turning the handle on washing machines. Now that more technology has come into homes, life has become even easier. We have vacuum cleaners that clean by themselves, bread makers that are completely automated, and clocks that set the time themselves. What fun it is to investigate these high-tech wonders!

Roomba Intelligent Floor Vac

Roomba is one of the first robotic devices that is truly useful in the home. It is the first stab at creating something like Rosie, Jetson's robotic maid.

Sandy's favorite

Roomba cleans quite well, using a "crop-circle" algorithm to cover the entire area. It picks up dust bunnies, dirt, and even dry cereal. It traverses the room in a large circular motion. It scurries against the walls and crisscrosses the area like it has a mind of its own.

If Roomba hits an object, it readjusts and continues on its way. The vacuum's low-to-the ground, flying-saucer shape allows it to easily get under furniture that might be difficult to reach with a conventional vacuum. Roomba is smart enough to avoid stairs. The two included battery-operated "virtual wall units" are used to block off doorways or openings up to 13-feet wide. In most situations, you will want to use these virtual walls to limit the robot to one room at a time because it works much better in a confined space.

Model: Roomba Discovery SE
Manufacturer: iRobot
URL: www.irobot.com
Price: $299.99

Every version of Roomba gets a little smarter. This Discovery SE version comes with a charging base. The vacuum senses when it needs to be recharged and automatically returns to its charging base. I can't wait to see what future versions can do!

iCEBOX CounterTop

This is the ultimate gadget for your kitchen. The iCEBOX CounterTop is a computer, television, and FM radio all in one. It supports dial-up Internet, broadband, and wireless broadband. You can attach a video cam to the iCEBOX and use it to monitor different areas of your home.

The touch-screen display is perfect for the kitchen. It is easy to wipe off, and there is no need to use a mouse. The wireless remote control and keyboard are waterproof. Wash them with mild detergent and rinse under the kitchen faucet.

Model: iCEBOX CounterTop
Manufacturer: Salton
URL: **www.esalton.com**
Price: $1,499.99

With the addition of the Smart Appliance Network Interface Card, the iCEBOX can be networked with other appliances in the Beyond line like the Beyond Bread Maker, the Beyond Microwave Oven, and the Beyond Coffee Maker. All are, of course, sold separately.

Beyond Bread Maker

The smell of fresh-baked bread brings memories of Grandma humming while her flour-covered hands gently kneaded the bread. Few people make bread by hand anymore, though. Now an electric bread maker and a box of bread mix give you fresh-baked bread quickly and easily.

The Beyond Bread Maker carries ease of use one step further than the average bread maker. It takes all the guesswork out of using a bread machine. You can't make a mistake because the machine actually determines exactly how to

Model: Beyond Bread Maker
Manufacturer: Salton
URL: **www.esalton.com**
Price: $149.99

prepare the bread. Just scan the UPC code from the package of bread mix, and the bread maker prepares it to perfection. Hundreds of barcodes are already in its memory, and you can also enter your own. The Beyond Bread Maker, when paired with the iCEBOX or the Beyond Connected Home Hub, can also learn new barcodes from the Internet.

The most work you will do is emptying the box and adding any other necessary ingredients. If you tell the bread maker what time you want the bread, you can get up in the morning or come home at night to fresh, perfectly baked bread with the touch of a button.

Emerson SmartSet

Who wants to have to reset the clock when the power goes off or when the time changes in the spring and the fall? Not me! That's why I have fallen for the Emerson SmartSet alarm clock. As soon as you plug it in, the SmartSet knows the correct time, date, and day of the week. All you have to do is tell it which time zone you are in. If the power goes off, the SmartSet clock resets itself automatically. It automatically adjusts for daylight savings time and leap year. This is all accomplished with a computer chip that is set with all the information needed to maintain the correct time, day, and date until the year 2096.

Model: CKS1850
Manufacturer: Emerson Research
URL: **www.emersonradio.com**
Price: $24.95

The blue LED clock display is large (almost 1 inch high) and easy to read. A dimmer switch allows you to lessen the intensity of the display. The dual alarm is easy to set with large buttons on the top of the radio. You can also set the alarms for every day, weekdays only, or weekends only. The sliders to turn the alarms on are inconveniently located on the side of the radio, but other than that, this clock radio is a dream to operate and to own.

Weather/All Hazard Radio

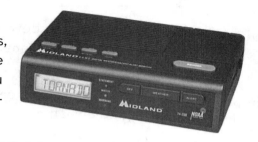

We can't control hurricanes, tornados, floods, or storms, but we can prepare for them. A weather radio can let you know when a storm or weather emergency is in your area. These types of radios have special receivers that pick up the 24/7 broadcasts issued by the National Weather Service. This Midland radio has a wonderful automatic alert system. You simply put the unit in a silent

Model: 74-200 S.A.M.E.
Manufacturer: Midland
URL: www.midlandradio.com
Price: $59.95

standby mode, and when an alert is issued that affects your area, the radio comes alive with a loud tone, a voice notification, or a flashing red LED alert. The name of the emergency scrolls across the LDC screen. It has 56 different alerts, including tornado, flood, blizzard, and high wind.

The 74-200 has an emergency battery backup system. It uses seven different National Weather Service channels that can provide coverage in 95% of the United States. It can receive broadcasts from the weather stations up to 50 miles away.

This little gadget is also a digital clock with alarm. You can listen at any time, day or night, to the latest weather information from the National Weather Service.

Dual Alarm Thermometer Clock

What will you wear today? The way you dress usually depends on the outside temperature. With this small device, you can wake up and know the exact outdoor temperature. The Thermometer Clock comes with two parts. The two-toned stainless and gray base has a large LCD

screen that shows both the indoor and outdoor temperature. The outdoor sensor is a
small white device that can be placed or hung
on the outside of the house. The base also
displays the time and day of the week. It sets
itself automatically, and because it synchronizes with the U.S. Atomic Clock, the time is
always accurate. You never have to reset the
clock because it automatically adjusts for
daylight savings time and leap years. The clock
runs on two AAA batteries, so
you don't even have to worry
about power outages.

Model: RMR602A
Manufacturer: Oregon Scientific
URL: www.oregonscientific.com
Price: $49.99

The attractive base can be wall
mounted or set on a desk or
nightstand. Both you and your
honey can use it because it has dual alarms.

Model: Sonic Boom Alarm Clock with
Bed Shaker
Distributor: Gold Violin
URL: www.goldviolin.com
Price: $44.95

Sonic Boom Alarm Clock

This is the perfect alarm for heavy
sleepers and the hearing impaired.
This clock radio emits a loud, pulsat-
ing audio alarm or shakes the bed to
wake you. If you are a really heavy
sleeper, you can use both. The shaking is done by a small pod that you
place between the mattress and the
box spring. The volume is adjustable,
and you also have control over the
tone.

The clock has a large, bright green
LED display and battery backup for
power outages. With the Sonic Boom
Alarm Clock, you'll never be late again.

Reizen Talking Atomic Alarm Clock

This clock uses radio signals to synchronize its time with the U.S. Atomic Clock in Colorado. The LCD screen shows the date, day, and time. Although the black-on-gray display is not as readable as some other digital clocks, you never have to look at the clock if you don't want to. The clock broadcasts the time in a pleasant female voice. You can even have the clock announce the time hourly between 8 a.m. and 10 p.m. You can turn this feature off if it doesn't appeal to you.

Model: Reizen Talking Atomic Alarm Clock
Distributor: Maxi-Aids
URL: www.maxiaids.com
Price: $35.95

The alarm sounds along with an announcement of the time. If you hit the Snooze button, the next beep and announcement are made in five minutes. If you are the type of person who hits the Snooze button too many times, the time announcement might be useful to jar you from your sleep.

Friedrich Air Cleaner

If dust, smoke, pollen, and airborne allergens make you miserable, it might be time to invest in an air cleaner. The C-90A contains a prefilter to catch larger particles, an electronic cell and collector plate to grab microscopic particles, and an activated carbon filter to remove most common odors and fumes.

At 19" × 15" × 19", this air cleaner is big and bulky. Although it's not extremely loud, you definitely can hear the fan whirling all the time. It has three settings: Low, Med, and Hi. Each one gets

Model: C-90A
Manufacturer: Friedrich
URL: www.friedrich.com
Price: $499

progressively louder. You won't want this cleaner set on Hi when you are trying to have a conversation.

Despite any drawbacks, this air cleaner is highly rated at performing its job. *Consumer Reports* lists the Friedrich C-90A as excellent at removing dust and pollen and very good at removing smoke. It also recommends the Friedrich because it emits less dangerous ozone than many other air cleaners.

Senseo Coffee Maker

Sandy's favorite

This is a kitchen appliance that is perfectly suited to Boomers and Zoomers. After the "empty nest" syndrome sets in, there is little reason to use a big coffee maker every day. The Senseo makes just one or two cups at a time. It uses premeasured pods and a 1460-Watt boiler to ensure just the right coffee taste.

The unique feature about this coffee maker is that it creates a foam coating on top of the coffee. While not actually cream, it does give the coffee a unique creamy quality.

Model: Senseo Coffee Maker
Manufacturer: Philips
URL: **www.philips.com**
Price: $89

The Senseo comes with coffee pods from Sara Lee/Douwe Egberts, which come in several different types like dark roast, Irish cream, and decaffeinated. These are packaged in foil pouches that each contains 16–18 pods. Pods from some other manufacturers come in individual foil containers, which help keep them fresher longer. You do have to choose pods that fit the Senseo, but you don't have to stick with the Sara Lee pods if you don't like the flavor or find that they don't stay fresh enough.

Sandy's Lingo List

The world of technology has created some crazy new words. Here are explanations for a few of the more unusual words used in this chapter.

identity theft—Stealing the identity of others by using their social security number, credit card numbers, and other personal information. The perpetrator can use your current accounts and open new ones.

light emitting diode (LED)—These displays are brighter than LCD and more highly colored, usually red or blue.

liquid crystal display (LCD)—Used in display screens for most portable computers and many small digital devices. These displays have two sheets of polarizing material separated by a liquid crystal solution. An electrical current passing through the liquid causes the crystal to align, allowing or preventing light from passing through.

U.S. atomic clock—Called NIST-F1 and created by the National Institute of Standards and Technology (NIST), the U.S. atomic clock resides in Colorado. It is capable of keeping time accurate to about 30 billionths of a second per year.

Chapter 9

Photography and Printing

We live in a society exquisitely dependent on science and technology, in which hardly anyone knows anything about science and technology.

—Carl Sagan

No gadget and gizmo book would be complete without a chapter on digital photography and printing. These are two areas where equipment has gotten cheaper while functionality has increased dramatically.

In 1994, I paid $500 for my first color inkjet printer. Today I can purchase a similar printer for less than $50. Digital cameras have also had dramatic price reductions over the years, while the quality of the equipment and the photos has improved significantly.

There has never been a better time to buy cameras, printers, scanners, and all the wonderful gadgets and gizmos that go along with them.

Digital Cameras

Digital photography has opened a whole new world to those who love to take pictures. Today's digital cameras are cost effective. They come in a large variety of sizes, shapes, and capabilities. All are feature-laden wonders.

Ever since I was a kid, I have been drawn to picture taking as a hobby, but I found standard photography quite expensive. Because digital cameras have no film, there is no additional cost for film or processing. As a result, I have been able to invest in a good digital camera and take all the pictures I want with no additional expense. That along with the instant results has been wonderful.

I'm working on composition and technique and investigating the many capabilities of my digital camera. Who knows? Someday I might even be an expert. The best part, though, is that I'm enjoying every minute. Why don't you join me in the brave new world of digital photography? Just look at these wonderful cameras...

Easy Share Camera with Printer Dock

This 10X optical zoom camera is part of Kodak's Easy Share high-zoom series. This 5-megapixel camera has a sturdy body and a nicely weighted feel. It won't fit in your pocket, but the shape allows you to wrap your hand around it for added stability. It is less than 4 inches wide—smaller than many other full-featured cameras.

Model: Z740 with Printer Dock Series 3
Manufacturer: Kodak
URL: **www.kodak.com**
Price: $479.95

This camera is easy for amateurs to use. It has 18 preset scene modes and 3 color program modes, including sepia and black and white. Yet you can set your own aperture, shutter speed, exposure compensation, and other settings if you wish. Although not a Single Lens Reflex camera, the included lens adapter allows you to mount an optional wide-angle lens or 55-mm filters. Controls are easy to use after you go through the manual to figure it all out. Although the screen is a little small, it is clear and easy to view. The Z740 has 32MB of internal memory and also accepts Secure Digital (SD) or MultiMedia Card (MMC) memory cards.

This camera is not called Easy Share for nothing. You can use a button on the back of the camera to tag photos to print or email later. Another nice feature is One Touch to Better Pictures, which automatically adjusts the settings on your inkjet printer to give you the best quality when printing on Kodak inkjet photo paper.

The printer is perfect for making 4" × 6" prints or printing smaller pictures on the 4" × 6" sheet. Just drop the camera in the dock, press the button, and you can print your photos without a computer. Or you can attach the printer to your computer and use it just like you would any other photo printer.

The whole unit is small enough to take to a birthday party or wedding, where you can surprise the guests with instant printed photos. Of course, you can also use the included USB cable and software to upload your photos to the computer and edit them before you print them.

Nikon Coolpix 8800

The Nikon 8800 is an 8-megapixel camera with a 10X optical zoom. Its special feature is the integrated Vibration Reduction, which automatically compensates for small camera movements that would otherwise cause the photo to blur. It's great for shaky hands. The integrated handgrip and weight of the camera are also useful for those, like me, who have trouble focusing the camera without moving.

The flip-out LCD screen is small but clear. Although there is no conventional optical viewfinder, you can use the LCD screen or the electronic viewfinder to compose your shots. Again, after you read the manual, the buttons and menu system are fairly easy to navigate. The camera uses a rechargeable lithium battery pack and a CompactFlash card. It comes with a remote control that you can use for self-portraits. The advanced user can set shutter speed, aperture, and other settings. The amateur can use the screen modes and automatic settings with great results.

You can use the included lens ring to attach optional lens adapters and filters. It also has a slot for an external flash. In addition, the Coolpix can record video sequences. The anti-shake feature really comes in handy within the video mode.

Model: Coolpix
Manufacturer: Nikon
URL: www.nikon.com
Price: $999

Sandy's favorite

Evolt

The Evolt is one of my favorite cameras. The quality of both the camera and the photos is quite remarkable. The Evolt has some unique special features. First of all, it is a Single Lens Reflex (SLR) camera, which means that the camera can use different lenses. There is a problem inherent to all digital SLR cameras. When the lenses are changed, dust can enter the camera and accumulate on the image sensor, causing spots and blurs on the photographs. Olympus has solved this problem with its innovative dust-reduction technology. I know professional photographers who have used this camera in dust-ridden places like rodeos without a problem.

The Evolt is the first digital camera that was built to be digital from the ground up. It uses

Model: E-300
Manufacturer: Olympus
URL: www.olympus.com
Price: $899

excellent Zuiko Digital Specific lenses. Other digital SLR cameras that use nondigital lenses can have a slight blurring around the edges of the lens because they were not created for digital imaging. The Zuiko lenses produce pictures that are sharp edge-to-edge.

With any high-end camera like this, reading the manual is a must, but menus and buttons are well designed and fairly intuitive. I like the Evolt for wanna-be photographers like myself. It has 14 scenes including portrait, fireworks, night scene, and sunset. You can use these to let the camera automatically adjust the settings when you don't feel like taking the time or when you just want to focus on your composition. When you have the time and are ready to experiment, you can set the white balance, aperture, and other settings yourself. The Evolt is also capable of taking photos in the RAW format. This format captures exactly what the camera sees, so it is useful if you want to play with the settings after you download the pictures to your computer.

This 8-megapixel camera takes great photos and might even propel you to become a great photographer.

PowerShot SD400

Although it can be exciting to use a big camera with plenty of features, there is nothing more exciting than pulling a small camera from your pocket to capture an event or scene that you otherwise would have missed. This Canon PowerShot is amazingly small for the power it packs.

This boxy, ultra compact, 5-megapixel camera measures in at just 3.4" × 2". With the camera at less than 1 inch thick, you can easily carry it in a shirt pocket or pocketbook.

I was amazed to find a large, clear 2-inch LCD display on such a small camera. When you're shooting in dark lighting, the LCD automatically brightens. The

Model: PowerShot Digital Elph SD400
Manufacturer: Canon
URL: www.canon.com
Price: $399

PowerShot also has a 3X optical zoom. I was also blown away by the quality of the photos when using the auto-adjust of the five scene modes. This camera really qualifies as a "point-and-shoot" camera. It also can record movie clips with sound. It even includes a compact movie setting that captures movies small enough to email. You can use the built-in microphone to attach a voice memo to the images.

Talk about special features! The PowerShot SD400 uses a new feature called My Colors Mode. This mode allows you to correct colors as you shoot. This camera can even take pictures underwater up to 10 feet deep with the addition of an all-weather case and an optional high power flash. The macro mode allows you to shoot larger-than-life close-ups with one button. It also has a useful optical viewfinder.

The PowerShot comes with a regular 16MB Secure Digital memory card. If you want to purchase a larger memory card, consider a high-speed memory card, like the SanDisk Ultra that is discussed in the "Photographic Aides" section.

This is a camera that won't weigh you down when you are on vacation. In fact, it's so small that you could carry it with you all the time.

Instant Replay Binoculars

Whether you want to get a close-up view of that sports event, take a picture of the kids playing soccer, or check out that last call by the umpire, these binoculars are for you.

Bushnell is known for its quality binoculars, and this pair lives up to the Bushnell name. Even at the maximum 8 × 32 magnification, the quality optics give a clean and clear view. These, however, are more than just binoculars. With the Instant Replay binoculars, you can take 3.1-megapixel still photographs. The CompactFlash card slot allows you to take as many photos as you like.

Model: 18-0833
Manufacturer: Bushnell
URL: www.bushnell.com
Price: $659.95

The instant replay mode handles up to 30 seconds of video. You can record these movies constantly so that the last 30 seconds of the action is always recorded. This uses quite a bit of battery power, but it is easy to keep a few spare AA batteries in your pocket when you use this function.

These binoculars are the only ones I've seen that actually take the picture through the binocular lens. This ensures that your picture will look exactly as you see it through the binoculars. There is a clear flip-up LCD screen for viewing photos and video. You can upload photos and movies to the computer via the USB cable. All-in-all, this is a wonderful pair of binoculars!

Photographic Aides

The variety of cool photographic aides is almost overwhelming. There are batteries, battery chargers, memory cards, memory card readers, drawing tablets, monitor calibration products, and even special cleaning devices.

Even if you don't want to invest in all these gadgets, it is always good to know that they are available when you need them.

Energizer Rechargeable Batteries and Charger

We all are concerned about the effect our lives will have on future generations. One way we can make a positive impact is to use rechargeable batteries.

Energizer has several compact chargers that can charge multiple batteries at one time. There are Energizer models that accommodate AAA, AA, C, D, and 9V batteries. They all work in a similar fashion.

The CHDCWB-4 that I looked at recharges AA batteries, the size most commonly used in digital cameras. The same charger can also accommodate AAA rechargeable batteries.

The new Energizer chargers are "more intelligent" than previous versions. They now have safety timers, temperature monitors, and trickle charge technology. The charger for AA/AAA batteries is a compact unit that folds to a smaller size when no batteries are inserted. Its size makes it suitable for traveling and for taking along with your camera equipment. It automatically detects which size of battery you have placed in it (AA or AAA). There is an indicator to show the status of the charge, and the plug flips down for easy storage/portability.

Model: CHDCWB-4
Manufacturer: Energizer
URL: **www.energizer.com/products/ rechargeables/sizes.aspx**
Price: Retail price $23.99, street price $17.99

If you need fast charging, look at the Energizer 15-minute charger. To charge your batteries on the go, you can get an Energizer car charger. If your camera takes regular batteries, you should always check the documentation to be sure that it can use rechargeable batteries. If it can, a battery charger like this might be just what you need.

RoadWired Pod

This compact carrying case is just 7 inches high, $4\frac{1}{2}$ inches deep, and $5\frac{1}{2}$ inches wide, but it has an amazing collection of individualized spaces. The Pod can easily handle a small or mid-sized camera, plus memory cards, floppy disks, PDA, batteries, pens, money, car keys, and more. It has a place for everything. The unique "winged" design allows for easy access to more than 20 pockets and compartments. Some sections are elastic; others have loops for cords, batteries, and adapters. Two outside mesh pockets hold often-used but easily lost components like lens caps and add-on cards.

The bag features quality construction and ingenious design. Velcro dividers keep cameras, PDAs, MP3 players, and other equipment from banging into each other. A special clear plastic-coated area inside the cover holds a business or identification card. There is even a secret pocket. You can carry the pod by the top handle, the shoulder strap, or on the belt loops. I love the red pod, but it also comes in black, navy, yellow, and olive. Podzilla is the larger $69.95 pod, and there are also two smaller pods ($29.95 and $24.95).

Model: RoadWired Pod, red
Manufacturer: RoadWired
URL: www.roadwired.com
Price: $49.95

15-in-1 Card Reader

What is the easiest way to transfer digital photos from your camera to your computer? You don't need to bother with drivers and software programs; you just need an inexpensive media reader. Hook it up to the computer with the attached USB cable, and the device appears as a removable drive under My Computer. Then drag and drop or copy the pictures to your hard drive. It's easy, and you don't wear down your camera's batteries during the transfer process.

The Belkin 15-in-1 reads and writes 15 different types of media cards including xD Picture Card, CompactFlash, SmartMedia, Memory Stick, Mini Secure Digital, and others. So if you are looking for the perfect present, you don't even have to know which type of media card your recipient's camera uses. The 15-in-1 Media Reader & Writer is also useful for copying or moving photos or files from one type of media card to another. Because it is a USB 2 device, the Belkin Reader & Writer transfers files quickly. If, however, you have a USB 1.1 port on your computer, the device still functions well, but it works more slowly.

Model: F5U249
Manufacturer: Belkin
URL: www.belkin.com
Price: $49.95

Single Slot Multi-Card Reader

If you use a Secure Digital (SD) card, Memory Stick, Memory Stick PRO, MultiMediaCard (MMC), or xD Picture Card, you can get a memory card reader that doubles as a portable USB hard drive. The Lexar Single Slot Multi-Card Reader functions just like a memory card reader. Insert your memory card into the reader and insert the reader into the USB port on your computer. The reader shows up as a drive icon on your computer, and you can easily move or copy the files to your computer. When you insert a memory card into the reader, it also functions like a USB hard drive that you can use to store files or move them to another computer.

This card reader is unique in that the memory card fits entirely in the reader and is protected by the cover when you use it to transport files. Not only is it unique, but it's also useful. You can use it for PCs and Macs.

Model: Single Slot Multi-Card Reader
Manufacturer: Lexar
URL: www.lexar.com/readers/ trio.html
Price: $22.99

Media Reader for iPod

If you are an avid photographer, you take a lot of photos, especially when you are on the road or at special events. What do you do when your memory card fills up? Well, if you have an iPod and the Belkin Media Reader for the iPod, it's easy. You simply attach the reader to the iPod and download all your photos to the iPod for storage. If you have a high-capacity iPod, it can hold thousands of digital photos. After you empty your storage card, you are ready to take more pictures, and your first batch is safe in your iPod. The Belkin reader transfers the photos by a speedy built-in connection. It supports CompactFlash, SmartMedia, Secure Digital (SD), Memory Stick, or MultiMediaCard (MMC).

Model: F8E461
Manufacturer: Belkin
URL: www.belkin.com
Price: $49.99

The Belkin Media Reader uses software that is already built into your iPod (software versions 2.1 or later), so you don't have to fumble around with settings.

Ultra II CompactFlash

All memory cards are not equal, especially if you have a high-megapixel digital camera. The SanDisk Ultra is a CompactFlash memory card with a minimum sustained write speed of 9MB per second and a read speed of 10MB per second. Don't worry about the specs—all you need to know is

that this card speeds up the data transfer in a high-end, high-megapixel camera. If you are using a camera like the Nikon 8800 or the Olympus Evolt, you will see faster shooting speeds and less time needed between shots, with this card. This Ultra card lets you save large images more quickly, so if you are shooting in RAW or bitmap, you will see a difference.

SanDisk Ultra cards are available in other formats, such as Memory Stick, Memory Stick Pro, and Secure Digital. Capacity of CompactFlash cards range from 64 MB to 4.0 GB.

Model: 512MB CompactFlash Card
Manufacturer: SanDisk
URL: www.sandisk.com
Price: $49.99

Graphire3 Tablet

Sometimes a mouse is just not good enough to get the detail you need on the screen. For many years, graphic designers have relied on an electronic tablet and pen to get the precision that they need. Because of their price, these tablets were previously used only by businesses and professionals. Wacom has changed all that with its new line of tablets that are affordably priced for the home market.

The Graphire3 hooks up to the computer via a USB connection. The package includes a tablet and an ergonomically designed cordless pen. Every point on the tablet corresponds to a point on the computer screen, so when you move your pen on the tablet, the cursor moves to the corresponding point on the screen. One of the best features of

tablets is that they are pressure sensitive. In drawing programs that support this feature, you simply press harder to get a larger brush stroke or a darker line. It is much easier and more intuitive than changing the size of the brush or the pen with your mouse. Paint Shop Pro, Photoshop Elements, Photoshop, and CorelDraw are just a few of the programs that support this pressure sensitivity. The Graphire3 comes with three excellent programs that you can use for drawing, painting, and designing your own pieces of art.

Model: Graphire3 6x8
Manufacturer: Wacom
URL: www.wacom.com
Price: $199.95

You can use the Wacom tablets with either PCs or Macs, and you can write directly in Microsoft Office XP and Apple Inkwell. Many other programs also support the tablet. You can use it to sign your name and annotate documents in Adobe Acrobat and Microsoft PowerPoint. The pen also has a pressure-sensitive eraser, just in case you make a mistake. Wacom includes a cordless mouse with its Graphire3 Tablet, giving you the option of using whichever tool is best for the task.

Spyder2 Plus

Sandy's favorite

Have you ever wondered why the colors you see on the screen are not the colors that print? It's because the monitor and the printer don't "talk" to each other regarding the colors they use. The solution is to calibrate the monitor to have it display accurate colors. Then the computer can send the correct colors to the printer.

The Spyder2 system does just that. It works with LCD, CRT, and notebook displays on both Windows and Mac computers. It even worked to

calibrate both of the monitors on my dual-display system. The calibration is straightforward. Install the software and attach the Spyder2 Colorimeter to the computer via USP port. Then simply follow the onscreen instructions to calibrate the monitor. The instructions include hanging the Colorimeter on the screen to do the color readings. When hanging on the screen, the Colorimeter looks like a large spider, hence the name, Spyder.

Model: Spyder2 Plus
Manufacturer: ColorVision by datacolor
URL: **www.colorvision.com**
Price: $269

After the calibration, you are shown a side-by-side comparison of the results, which are often dramatic. The entire process takes about 30 minutes, and it is recommended that you do this every 2–4 weeks because the monitor's colors can shift over time.

The Spyder2Plus combines the monitor calibration with printer color correction and Adobe Photoshop Elements 3.0. A home version with a little less functionality and without the Adobe software is available for $119. The Spyder2 works on Windows 2000 or XP and Mac OS 10.2 or better. If you want to control the color on your monitor, this is the product to do it!

Oxyride Extreme Power Batteries

The New Oxyride Extreme Power batteries are said to last up to two times longer than traditional alkaline batteries in digital cameras and 1.5 times longer in other electronic equipment. Panasonic calls this the most significant development in primary battery technology in the past 40 years— the biggest breakthrough since alkaline batteries.

The Oxyride batteries use new materials and advanced manufacturing technology to incorporate a higher quantity of electrolyte in

Model: 4-pack AA
Manufacturer: Panasonic
URL: **www.panasonic.com/consumer_
electronics/batteries/oxyride.asp**
Price: 4-pack AA or 4-pack AAA, $3.99

the battery. Panasonic says this gives the batteries more power (1.7 volts compared to the 1.5 volts of normal alkalines), resulting in more muscle for battery-powered equipment. I tested them using two identical flashlights, one with alkaline batteries and one with Oxyrides. The flashlight with the Oxyride batteries produced a dramatically stronger, brighter light.

In just one year of availability in Japan, Oxyride batteries have captured 10% of the Japanese battery market. They hit the American market in June 2005 in both AA and AAA sizes. Believe it or not, they are slightly cheaper than premium alkaline batteries.

R.A.P.S.! Advanced Protection System

The R.A.P.S.! has an Advanced Protection System built into it that offers high-tech protection against environmental pollutants and corrosion and makes this product truly unique. This wrap has a multilayered construction using a proprietary material called Corrosion Intercept that was originally designed to protect missile and space components. This material keeps gases and other environmental corrosives from damaging the contacts, switches, housings, and circuitry components in cameras, computers, and other electronic gadgets.

Model: R.A.P.S.! Advanced Protection System, Large
Manufacturer: RoadWired
URL: www.roadwired.com
Price: Large $17.95, Small $12.95

The wrap is a flat multilayered fabric that can be draped around any equipment that you want to protect. It's perfect for cameras. The Velcro corners hold the R.A.P.S.! in place. It is ideal for added protection when you are packing your equipment inside luggage, a carry-on, or any type of case. You can also fold the wrap or roll it into any configuration to conform perfectly to the piece of equipment that you want to protect.

The wrap comes in two sizes: 16" × 16" and 20" × 20". Both sizes come in black, gray, yellow, or red.

Norazza Digital Media Slot Cleaners

You paid a lot of money for that digital camera, so you'd be smart to take good care of it. Your mother told you that cleanliness was important, but you probably always found it difficult to keep everything clean. Well, digital cleaning kits make it easy by putting everything you need in one place. They are designed for digital personal electronics with memory card slots (such as PDAs, digital cameras, flash card readers, MP3 players, and photo printers).

You purchase the proper kit for the type of memory card you are using. Kits are available for CompactFlash, Memory Stick, Secure Digital (SD), MultiMedia Card (MMC), and xD Picture Card. The cleaning card fits in the memory slot, just like any regular digital media card. It cleans and polishes the connectors inside the digital media slot. You insert the card as you would a normal digital media card and then repeat the process 2–3 times. There is even a cleaning cloth included with which you can dry and polish the monitor, lens, and surface of your digital camera or other digital device as a final step in cleaning. Be sure to put the cleaning cloth in your camera bag. A clean lens takes a better picture.

Models: C225E & C200E
(CompactFlash and Digital Card)
Manufacturer: Norazza
URL: **www.norazza.com**
Price: $19.99

Print, Scan, and Display Your Photos

After you've captured your masterpieces, you will want to show them off to family and friends. You can print them, create a slide show on your television, or display them in a digital picture frame. These new gadgets let you enjoy working with your digital photos. New technologies have created better printers that produce excellent quality photo prints. Scanners let you scan in your old photos, slides, and film strips to enhance and preserve them. The variety of digital picture frames is amazing. All these gadgets and gizmos bring out the best in your photos and let you have fun doing it.

Photosmart 375B

Wouldn't it be great to be able to share your prints with others right after they were taken? You can with the Photosmart 375B. It was made for the photographer on the go. This printer is compact and weighs less than 3 pounds, making it very portable. Although this printer looks somewhat like a toaster, it performs like a first-class photo printer. It prints 4 × 6" prints quickly and easily. You can also print two 4 × 3" photos or four 2 × 3" photos on a single sheet of photo paper. The 375B can even print still photos from a video.

Just slide your memory card into the printer to print directly from the card. The printer has slots for all the major memory cards, including CompactFlash Type I

> **Model:** HP Photosmart 375B
> **Manufacturer:** Hewlett-Packard
> **URL: www.hp.com**
> **Price: $249.99**

and II, Memory Stick, Secure Digital, MultiMediaCard, SmartMedia, and xD Picture Card. You can also print directly from any PictBridge camera (see the Lingo List at the end of this chapter). The 2.5-inch flip-up display makes it easy to preview photos before you print. When you transport the printer, the display folds down for protection. You can even perform basic editing like rotating, cropping, and fixing red eye. You can delete pictures from the memory card right from the printer.

This printer begs to be taken on the road. It comes with a rechargeable battery back. You can also power it via a car charger or a wall socket. The 375B works with PC or Mac. Take it on vacation, and you can send your friends instant custom postcards!

Pixma Photo Printer

In recent years, Canon has dramatically improved the quality of its inkjet printers. That, along with some of the best printing software in the industry, makes Canon a good choice for an inkjet printer. The Pixma is no exception. Its 8-color ChromaPLUS ink system produces professional results. Although there are more ink cartridges to replace, this printer uses individual cartridges, so you only have to replace the color you need.

With the Pixma, you can often print without a computer. Any PictBridge (see Lingo List at the end of this chapter)-enabled digital camera or camcorder hooks directly to the printer. The Pixma is fast. A 4" × 6" print takes only about 21 seconds to print.

Model: iP8500
Manufacturer: Canon
URL: **www.canon.com**
Price: $349.99

I like the low-ink reminders that give you enough notice so that you can purchase more ink. I also love the Canon software. The Pixma comes with Easy-PhotoPrint 3.0 (Windows/Mac), PhotoRecord (Windows), and Easy-WebPrint (Windows only, requires Microsoft Internet Explorer version 5.5 or later). This software makes it easy to print different size photos and automatically formats web pages for proper printing without wasting paper or cutting off the edges of the print.

HP Deskjet Color Inkjet Printer

If you print a lot of photos, you definitely should purchase a dedicated photo printer. If, however, you want a printer that will do double duty for photos and everyday printing, this Deskjet printer might be just what you need.

The printer attaches to a PC or Mac with a USB cable and is easy to set up. The 17.7" × 14.5" footprint is an average size for an inkjet printer.

At 30 pages per minute in black and 20 pages per minute in color, it is speedier than most other inkjets. The text output is good, and photos quality is also at the high end. This Deskjet takes two cartridges. The only downside is that you need to change the print cartridges choosing black, photo color, or photo gray as appropriate for the type of printing that you will be

Model: 6540
Manufacturer: Hewlett Packard
URL: www.hp.com
Price: $129.99

doing. On the plus side, if you often print photos in black and white, the photo gray cartridge will be useful.

If you are looking for a good all-around printer, this might be what you're looking for. Not too many out there have both the speed and quality that this one does.

Epson Perfection Photo Scanner

Over the years, I had accumulated boxes and boxes of old photos, slides, and negatives. Some of my photos are faded and scratched; some of the negatives are slightly damaged. All are totally disorganized. I decided it was time to embark on a project to get control of my old photos.

The Epson Perfection Photo Scanner was just the tool I needed for the job. With the built-in transparency unit, it accommodated all my film. It also handles multiple slides simultane-

Model: Epson Perfection 4990 PHOTO
Manufacturer: Epson
URL: www.epson.com
Price: $449.99

ously. This unit can accommodate 35mm slides, 35mm strips (as many as 24 frames), medium-format film, and 4" × 5" film. Although this scanner's rounded metallic gray case looks chunky, it has a relatively small footprint.

Hook up the scanner by USB or the faster FireWire connection, if you have it available. Hardware and software installation is easy. The scanner comes with plenty of software. I especially like the defect removal capability of the built-in software. Although it increased the time needed for scanning, dust and scratches were automatically removed, and colors were restored.

The auto mode is wonderful. In this mode, the software recognizes the document type and performs the scan. If you have more than one photo in the scanner, it scans each one separately. It also straightens any photos or documents that are not exactly square. Of course, you can choose to control all the scanning settings yourself. The 4800 × 9600 dpi resolution and 48-bit color produce replication of even intricate details in photos.

Ceiva Digital Photo Receiver

This might better be described as a picture frame, because that's what it looks like. The Ceiva looks like an ordinary 8" × 10" wood picture frame with traditional black matting covered by glass. However, it does much, much more than your mother's old picture frames.

With the Ceiva, you can upload pictures to the Internet and have them automatically appear in the picture frame. Pictures are transferred to the frame via telephone connection, so the person who uses the frame doesn't need to have a computer. Others can post the pictures and have them appear in the frame. So you can give great grandma the Ceiva frame. Then family members around the world can post their pictures through the Internet. The picture frame automatically calls the Ceiva website during the night and downloads the photos to the frame. When grandma gets up in the morning, she sees all the new pictures.

Model: LF-2003
Manufacturer: Ceiva
URL: **www.ceiva.com**
Price: **$139.95, plus $9.95 per month (check Ceiva site for rebates/special offers)**

The display is approximately 5" × 7" with a resolution of 640 × 480 VGA. Ceiva's approach is thorough and well implemented, but more importantly, it's simple. Everything is adjustable. The frame can be turned on and off automatically at certain times of day; the telephone access number can be changed; and the slide show interval can be adjusted. You can do all this on the Internet or by calling the support number.

The frame has just two buttons; one button adjusts the brightness, and the other controls the pictures. To cycle through the pictures, press the button and release. Pressing the button and holding it down starts an automatic slide show of the pictures. To connect to the Internet for new pictures, simply press and hold the button down longer until instructions appear in the frame telling you when to release. Or let the pictures download automatically during the night. It couldn't be easier.

The Ceiva frame holds 20 pictures. You can also keep up to 1,000 pictures in your Ceiva Internet account and can change the images as often as you want. The digital pictures that appear when the frame is plugged in make an excellent presentation. You can adjust brightness with a simple button on the back of the frame.

The Ceiva frame is unique, novel, useful, and easy to use. It would be a great present for any technical holdouts on your gift list, or it would be a perfect novelty for the desktop of any high-tech type. The only drawback is the monthly fee.

MemoryFrame

The MemoryFrame 810S is an 8" × 10" picture frame that can show off all your digital photos. Setup is easy. You can copy photos to the picture frame in several different ways. A digital camera can be connected directly to the picture frame with a USB cable. You can also transfer pictures from a card reader or a USB hard drive that is attached to the frame's USB port. Last but not least, you can install the included software on your PC and transfer the photos directly from the PC. If you choose this last option, you can use the included Digital Pix Master software to create entire slide shows that you can download to the frame. Pacific Digital also has a wireless picture frame that can talk directly to your wireless network.

Even if you don't have a computer, you can still use the MemoryFrame. You can transfer the photos directly from a camera or USB card reader, and the

Model: MF810S
Manufacturer: Pacific Digital
URL: www.pacificdigitalcorp.com
Price: $449.99

frame automatically creates a slide show. So you can use this frame as a present to great grandma or someone who doesn't have a computer. Change the pictures by simply connecting to your camera or USB memory card reader.

After your slideshow is in the frame, you can change the timing, transitions, and other slide show options or delete images by using the buttons on the back of the frame. You can also use buttons on the frame to manually switch to the next image. There are also pause and reverse buttons, a brightness control, and a volume control.

One nice feature about this frame is that you can easily remove the frame around the picture screen and add any standard 8" × 10" frame of your choice. The other great feature is the absence of monthly fees.

VistaFrame

This is one of the lowest-priced digital picture frames that offers good quality photos with no monthly fee. Unlike the MemoryFrame, which can use regular picture frames, and the Ceiva, which looks like it is in a wooden frame, the VistaFrame looks high-tech. The 6.8-inch LCS screen is surrounded by a satin-finished steel-looking bezel that is wider on the bottom. It has a small curved stand. There is no mistaking that this is a digital picture frame.

Besides being the cheapest of the digital frames, it is also the easiest to set up. Take the memory card from your camera and insert it into the frame. Press a few buttons on the side of the frame to copy the pictures to the frame. Then choose whether you want to show a single picture or create a slide show. The VistaFrame has

Model: VistaFrame
Manufacturer: Vialta
URL: www.vialta.com
Price: $249.99

two built-in memory card readers that accommodate five different memory card formats: Secure Digital, Memory Stick, SmartMedia, MultiMediaCard, and CompactFlash.

One cool thing about the VistaFrame is that it goes into a power-saving mode after 60 minutes of inactivity. A built-in motion sensor wakes it up when it detects movement in front of it, when a memory card is inserted, or when a button is pressed.

Radiant Frame

The Radiant Frame is not really a digital picture frame. Instead, it is a low-tech picture frame that displays digital photos that have been printed on an inkjet printer. The Radiant Frame is both unique and inexpensive.

Three different Radiant Frame styles are available: a wood frame with natural oak finish, a wood frame in matte black finish, and a glass frame in a somewhat contemporary design. The frames are approximately 8" × 10". They come with matting that accommodates a 4" × 6" picture. The frame itself is backlit, powered by either the included AC power adapter or a set of four AA batteries.

Model: Radiant Frame
Manufacturer: Radiant Frames
URL: www.radiantframes.com
Price: $29.99

The Radiant Frame comes with several sheets of special inkjet printer paper. This paper looks and feels like a translucent parchment-type paper. First you choose the photo you want to frame. Then you print your photo on the special paper. You insert the photo into the frame just as you would insert it into any ordinary picture frame. Turn the Radiant Frame on, and you are done. The paper is easy to print on, and the entire process is quite simple.

The combination of the special paper and the back light really make the picture pop. Although garden and landscapes look good, the frame really brings out the best in pictures featuring people. It makes them look much more lifelike. By the way, the nice glow of the Radiant Frame makes a great nightlight.

SanDisk Photo Album

The SanDisk Photo Album is small, compact, and lightweight and has a sleek look. You can view, store, and share your favorite digital photos, video clips, and music on your TV. No computer is required! It's small enough to take with you anywhere and can be used wherever there is a TV. You can choose to have the display in English, French, Spanish, German, Italian, Chinese, or Japanese, making it a great gift for those in other countries. It also switches between NTXC and PAL TV format, allowing it to work with any TV standard in the world.

The SanDisk Photo Album is simple, convenient, and affordable to use. You can view your images instantly as thumbnails or full screen by inserting the memory card from your digital camera into the SanDisk Photo Album. Eight different types of memory cards are supported: CompactFlash Type I and II, Memory Stick, Memory Stick PRO, SmartMedia (SM), xD Picture Card, Secure Digital (SD), and MultiMediaCard (MMC).

Model: SDV2-A-A30
Manufacturer: SanDisk
URL: www.sandisk.com
Price: $49.99

Use the remote control to select transition effects, to rotate and zoom in/out on a photo, to expand the video to fit the entire screen, and more. Simply press the Store button on your favorite images, and the SanDisk Photo Album instantly creates a photo album. On the back is a slot for a second CompactFlash card (not included), which is used as the "memory" for storing slide shows, video clips, and MP3 music selections. The storage format for photos is optimized for television viewing. You can also drop MP3 music files onto the same card that stores the images for TV playback and thus have background music for slide shows.

PhotoBridge HD

The Roku PhotoBridge is an excellent digital media hub. The unit itself looks like a small piece of audio equipment. You need to read the instructions, but setup is straightforward. After you've attached the unit to your television, you can insert the memory card from you camera and

see the photos displayed on your television. If you happen to have a high-definition TV, the Roku makes your photos look especially stunning. You can pause, skip, zoom, and rotate images at the click of a button.

Model: HD1500
Manufacturer: Roku
URL: www.rokulabs.com
Price: $399

You can hook up the PhotoBridge to your computer to show video saved on your computer. If you have a computer that can record television, this unit allows you to play your recorded television on a large screen. You can easily add the Roku to either a wired or a wireless network. It also allows you to play digital music in many different formats.

The HD1500 model includes custom Roku art packs that can turn your television screen into a wonderful display. The art packs include classic pictures from Monet, Picasso, Renoir, and others. They can also turn your TV screen into a life-like aquarium. Other art packs are available to suit your fancy.

Sandy's Lingo List

The world of technology has created some crazy new words. Here are explanations for a few of the more unusual words used in this chapter.

FireWire—A type of standard also known as IEEE 1394. FireWire is faster than USB. It supports data transfer rates of up to 400Mbps. Because of its high speed, FireWire is often used for transferring video files. All Apple computers have FireWire ports, but only some PCs have them.

inkjet printers—These printers use drops of ink to produce an image on the paper. Low-end inkjets use three ink colors. Higher-end printers use as many as ten.

megapixel—Digital imaging devices measure resolution by megapixel, which is roughly the equivalent of one million pixels. A pixel is the basic unit of composition used in an image. It is often thought of as a dot. A 5-megapixel digital camera would have a picture composed of more than 5 million color dots (pixels).

memory cards—Also referred to as media cards or storage cards. They include many different types, including CompactFlash, SmartMedia, Secure Digital (SD), Memory Stick, MultiMediaCard (MMC), and xD Picture Card. These memory cards are different sizes. The smallest is about the size of a postage stamp. The largest is slightly smaller than a book of matches. They are not interchangeable.

PictBridge—A technology that lets you print directly from your digital camera to a printer without using a computer. The camera and the printer can be from different manufacturers, but they both must support the PictBridge standard.

point-and-shoot—A term used for cameras that have automatic settings that allow the user to simply point the camera at an intended subject and press a button to shoot the picture.

RAW—In computer terms, RAW means untouched or original. A RAW file created by a digital camera contains the actual data captured by the camera's sensors without compression or changes.

resolution—The degree of sharpness of a character or image that is displayed or created. The term *resolution* is used for many digital devices, including computer monitors, printers, scanners, and cameras.

Single Lens Reflex (SLR)—This type of camera can use different lenses. An SLR camera uses a technology that allows the camera user to determine what parts of the image are in focus by looking through the lens. While the LCD screen on many digital cameras is used to frame the picture, on an SLR camera, the photographer looks through the lens to see the image. The LCD is used for reviewing the pictures and changing settings.

Chapter 10

Travel and Automotive

For my part, I travel not to go anywhere, but to go. I travel for travel's sake. The great affair is to move.

—Robert Louis Stevenson

Computers and technology have changed the way we travel. From new modes of transportation like the Segway to small devices that make travel easier, gadgets and gizmos abound for every need.

The automotive world has also undergone a technological revolution. Cars today have more computerized parts than ever before. Numerous wonderful add-on devices are available for automobiles. Whether it is a talking tire gauge or a device to help you get out of the car more easily, you can find many automotive gadgets to whet your appetite.

Navigation Devices

My husband always seems to be able to find anything. He's great at directions and never gets lost. I'm the total opposite. I never know whether I should turn left or right. Although my hubby might never need a global positioning system (GPS), these wonderful gadgets have become a welcome addition to my arsenal of high-tech gadgets.

We have no excuse for getting lost anymore. Technology has made it possible for us to know exactly where we are and where we are going. GPSs are able to bounce a signal between a network of satellites and a GPS device to accurately determine the position of the GPS device. This technology spawned the development of many different types of useful GPS mapping devices. When it comes to navigation, even the old-fashioned compass has gotten a digital facelift.

Tom Tom Go

The TomTom Go is an amazing device. This 4.5" × 3.6" × 2.3" unit comes with a small base that attaches to the car's front window with a sturdy suction cup. Just

press the lever to attach the suction cup, insert the power cable into the cigarette lighter, turn the TomTom on, and you are ready to go. This

TomTom successfully accesses the satellite system in less than a minute. This is an important feature because some of the GPS devices I tried took as long as 10 minutes to determine my location.

The TomTom is operated by a convenient touch screen. When you enter an address, TomTom creates a route from your location to that address. A map appears on the screen. You have several different map configurations to choose from. The best part, however, is that the

Model: TomTom Go 300
Manufacturer: TomTom
URL: **www.tomtom.com**
Price: $699

TomTom gives you verbal turn-by-turn instructions. The screen shows you where you are and the voice tells you exactly where to turn with just enough advance notice. You can choose from many different voices, including an Australian lady and an American man. TomTom has 50 pre-programmed voices and speaks 30 different languages.

When you plan your route, you can choose the fastest, shortest, or simplest route. You can also avoid certain areas or routes, such as toll roads. As you drive, the TomTom also locates churches, airports, restaurants, gas stations, and other points of interest. You can even have it give you an audible alarm when you come upon a point of interest.

I love the TomTom Go. It makes me feel just as directionally capable as my husband, and it has no monthly charges.

Streets & Trips

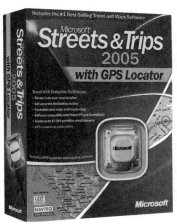

Streets & Trips has always been a good mapping program, but the addition of a GPS and with GPS capabilities programmed into the software, it's even better.

The software alone gives you a detailed route map and driving instructions. You can have the software show you points of interest, restaurants, gas stations, and ATMs along the way. Just put in your starting point and your destination, and you are given detailed directions. Add other information like your gas mileage, and the program spits out gasoline costs and other pertinent information. You can customize your trip to suit your needs. You can use the Streets & Trips software to plan your trip at home and print out the maps and driving instructions.

Model: Streets & Trips 2005
Manufacturer: Microsoft
URL: www.microsoft.com/streets
Price: $129.99 with GPS, $39.95 without

Another option is to install Streets & Trips on a laptop computer, plan your trip, and take the laptop along for the ride. You simply plug the small global positioning unit into the USB port of your laptop. You can place the GPS device on the car window with the included suction cup. Then watch your progress as you drive. A small car appears on the computer screen indicating exactly where you are and how you're progressing on your trip. If you make a wrong turn, the program recalculates your route automatically. As you drive, you can see the restaurants, gas stations, and other points of interest right on your computer screen. With this GPS system, your travel companion will be able to easily say, "Are you hungry, dear? There is a McDonald's coming up on the right."

You can purchase the Streets & Trips program with or without the GPS unit. No monthly fee is involved. The GPS functionality in the program works with any GPS device.

A Pocket PC version of Streets & Trips is included in the package. You need to purchase an adapter to attach the GPS unit to your PDA. After you have your PDA set up with Streets & Trips, it is good for walking trips in a strange area or for use in the car.

This travel system has a few small drawbacks. You must have a navigator in the car with you because you cannot safely look at the computer screen and drive at the same time. On long trips, your laptop battery might not last as long as you want to drive. The Streets and Trips program has no audio instructions.

DeLorme Earthmate® GPS

This is another software and hardware combination that turns your laptop into a GPS tracking device. The Earthmate GPS receiver connects to your laptop by a USB connection. Like the Streets & Trips solution, you load the included software on your laptop, attach the GPS to the front windshield with the suction cup, and connect it to your laptop with its USB cable. Click on the GPS tab in the software, and you can see your position in real time as you drive. It's pretty cool to see your vehicle travel on the screen as you travel on the road.

DeLorme pioneered the GPS for laptops concept in the mid-1990s. Because DeLorme developed both the software and the GPS, they work seamlessly together. The software maps the route for you and displays it on your screen. It gives you a detailed map and driving instructions. It also displays points of interest.

The Earthmate GPS and Street Atlas have one feature that Streets & Trips with GPS doesn't have. The Street Atlas program gives you voice prompts. The voice is computerized and a bit stilted, but it tells you exactly where to turn. Although you might still want to have another person in the car to navigate, these voice prompts make the system more fun and more useable. The ultimate coolness is exhibited when you instruct the GPS to search for services like restaurants near your position, and the voice reports a list of pizza places, Chinese restaurants, and others in the surrounding area. The Earthmate system has no monthly charge.

Model: Earthmate® GPS LT-20 with Street Atlas USA
Manufacturer: DeLorme
URL: www.delorme.com
Price: $99.95

Talking Digital Compass

Okay, I admit it. I get lost easily. And what gadget can always come in handy when you don't know which way to turn? A good old-fashioned compass. Yet this compass has a few high-tech twists. It has advanced magnetoresistive compass technology with no moving parts, so it is always accurate. Unlike a compass that has an arrow, this one has a color display. If you are hard of hearing or just love high-tech gadgets, you'll like the fact that this compass talks to you. And one smart compass it is—it speaks multiple languages!

Model: Maxi-Aids Item #C2000
Manufacturer: Maxi-Aids
URL: www.maxiaids.com
Price: $59.95

Available in English and Spanish or English and French, this gadget is powered by two 12-volt batteries.

Automotive Gadgets

Plenty of things can make your motoring just a little better. You might find that you need a bigger rearview mirror, a place to put your cell phone, or a way to get out of the car a little more easily. Don't worry; I've got you covered with this wonderful assortment of gadgets and gizmos for your car.

Sit-N-Lift Power Seat

For those who have disabilities, this power seat can be a valuable gadget. Sit-N-Lift is a fully motor- ized, rotating lift-and-lower passen- ger seat. It fits in the back passenger seat and completely lifts and rotates the passenger. Yet the seat looks like the other seats and fits without altering the vehicle. You can operate the seat with a remote control. For added safety, you can operate the seat only when the door is open and the vehicle is in park.

Product image not available for reproduction.

See vendor website.

Model: Sit-N-Lift
Manufacturer: GM
URL: www.gm.com/automotive/ vehicle_shopping/gm_mobility/ vaa_snl.html
Price: Varies

The beauty of the Sit-N-Lift is that it takes the passenger from the seated position in the car to the outside of the car and positions the seat and the passenger just 20 inches from the ground.

Sit-N-Lift is a dealer-installed option that is available on select GM sport vans and extended length minivans. The cost can often be wrapped into the financing when you purchase a new vehicle.

Park-Zone Platinum

If any of you have a tennis ball hanging in your garage to tell you when you have driven far enough into the garage to park the car, you can throw that old ball in the garbage. This is the high-tech equivalent.

Park-Zone is a great gadget for people who have a big car and a small garage, or in my case, a small car and an overcrowded garage. Park-Zone has an advanced microcontroller that handles the timing. It works like a traffic light. When you drive into the garage, a green light illuminates. A yellow light warns that you are approaching the park zone. A red light comes on when it is time to stop.

Model: PZ-1500
Manufacturer: Measurement Specialties
URL: www.msiusa.com
Price: $29.99

The platinum edition comes with an AC adapter, but the unit can also work with four AA batteries. Although the Park-Zone must be attached to the wall with four screws, installation is easy. After it's installed, you park your car in the garage in the proper position. Park-Zone uses ultrasonic sensing to determine your park zone by setting the precise distance from your bumper to the garage wall. After this one-time setup, the Park-Zone gives you the red light at exactly the right moment.

Talking Tire Gauge

You will get better gas mileage and a smoother ride when your tires have the right amount of air pressure. Yet, who wants to check the tire pressure? Well, with this cool device, you will definitely be more inclined to do that.

The Talking Tire Gauge has a nice handgrip, so it is easy to handle. Just align the nozzle on the tire gauge with the valve stem on the tire to get the pressure reading. The reading is digitally displayed on the small, clear screen, but the cool part is that the measurement is also spoken in a clear computerized voice.

Model: MS-4440
Manufacturer: Measurement Specialties
URL: www.msiusa.com
Price: $14.95

Even after the gauge has shut off, it remembers the last reading. Just press the Recall button to have the gauge announce the last measured pressure.

Driving Mirror

Do you ever wish you could see more in the rearview mirror? Here's a simple little gadget that does just that. A precision optically designed mirror fits over the rearview mirror with a sturdy clip. It increases your peripheral vision and enlarges the viewing area dramatically. The $10" \times 2\frac{1}{4}"$ mirror gives you greater vision and more confidence when backing up and viewing cars and other objects in the rearview mirror.

Model: Driving Mirror Item #70330
Manufacturer: Maxi-Aids
URL: www.maxiaids.com
Price: $24.75

Swivel Seat Cushion

If you or someone you know has trouble getting in and out of the car, this little gadget makes that task a breeze. The swivel seat cushion is a thick round poly-foam pad on a flexible plastic swivel base. The pad is covered with a removable, machine-washable cover. Just put it in the car and sit down on it. The base turns 360 degrees, so when it is time to get out of the car, there is no strain on the back or hip. You simply swivel to the side and get up. Sometimes the simplest devices are the ones that work the best!

Model: Swivel Seat Cushion Item #CU024
Manufacturer: Gold Violin
URL: www.goldviolin.com
Price: $24.95

Super Size Sticky Pad

This is another simple product that serves a need. The Super Size Sticky Pad is 8 3/4" × 6 7/8" × 3/16". Installation couldn't be easier. Take it out of the package, and put it on your dashboard. It clings to the dashboard without adhesives, so it doesn't mess your car. If you want to reposition it, just lift it up and put it down where you want it. It is washable and temperature resistant. I left one on my dashboard while my car was parked in the hot Carolina sun for several weeks with no ill effects.

The purpose of the sticky pad is to hold objects in place on your dashboard. You can use it for most anything, but I find it a great place to put my cell phone so I can answer easily if the phone rings while I'm driving. It is also super for keys, PDAs, and music players. For a while, I had a small bobble-head dog sitting on my Sticky Pad. My grandson and all the kids in the neighborhood loved it!

Model: Super Size Sticky Pad
Manufacturer: Handstands
URL: **www.handstands.com**
Price: $9.99

Avoid using the Sticky Pad on a hard or painted plastic dash, but it works great on most cars. If the stickiness wear off, you can renew it by washing with soap and water, but I have been using mine for more than a year and it still sticks perfectly.

The super size pad is available only in black. A smaller 6 3/4" × 4" pad comes in black, red, blue, or white and is available for $8.99. This is one great product for the car. If you have anything that you want to keep in place in your home, it could be useful in the house, too.

Griffin iTrip FM Transmitter

After you have all your digital music on your iPod, you probably want to listen to it everywhere. Yet, it is not wise to use headphones or earbuds when you are driving. iTrip has the solution. Just pop the small cylinder-shaped iTrip on top of the iPod. The iTrip is powered by the iPod. iTrip turns on automatically when you attach it to the iPod.

When it's attached, you can listen to music from the iPod through any FM radio. You choose any radio station on the dial for the best possible performance. You do this by choosing special station codes from the iPod. Although the iTrip is perfect for travel, you can use it with any FM radio. That features makes it great for using at home or in the office, too.

Model: iTrip
Manufacturer: Griffin Technology
URL: **www.griffintechnology.com**
Price: $35

Special iTrips are available that fit the iPod mini and that match the black U2 iPod.

Mobile Cassette Adapter

Although the iTrip is a great way to listen to an iPod on the road, other solutions allow you to listen to any digital music player when you are in the car. If you have a cassette player in your car, the Belkin cassette adapter is the perfect solution. This device looks like a cassette tape with a wire and plug attached. You simply put the cassette in your cassette player and plug the adapter into your digital music player. You can also use it in a portable tape player to play your digital music at the beach, while camping, or anywhere else you might listen to a portable cassette player. The Belkin adapter reduces noise by automatically maintaining the correct tension within the cassette deck as you drive.

Model: Belkin Part # F8V366
Manufacturer: Belkin
URL: **www.belkin.com**
Price: $24.95

Travel Helpers

Would you like a new high-tech way to get around? Or are you just looking for a little something to lean on when you walk? These devices also help you wake up when you are on the road and stay organized when you travel.

Segway Human Transporter

If you want to travel in high-tech style, the Segway is the way to go! The Segway, a self-balancing human transporter that

Sandy's favorite

looks like a scooter, is really breakthrough technology. It responds to your body in the same way that your own arms and legs respond to your mind. It's not your daddy's bicycle, for sure!

I was amazed to find that it only took a minute or two to learn how to ride the Segway. Lean ever so slightly forward, and the Segway moves forward. Lean back, and it moves back. Stand straight, and the Segway stops. To turn left or right, you simply turn the handlebar. Although I was only able to try the Segway in a structured, level area, it performed admirably. It is not only easy to use, but fun.

Model: Segway Human Transporter p Series
Manufacturer: Segway
URL: www.segway.com
Price: $3,999

The Segway is sturdy and obviously made to be durable. Solid state electrical systems and state-of-the-art motors ensure longevity, and the modular design makes it easy to service when necessary. Two rechargeable nickel metal hydride batteries power the Segway. Under ideal conditions, the Segway can travel up to 15 miles on one charge. The wheels can rotate in opposite directions, enabling the machine to turn in place. Five gyroscopes and tilt sensors monitor the rider's center of gravity about 100 times per second, keeping the rider balanced at all times. The Segway has three keys that the rider can choose depending on riding environment and level of experience. The Learning Mode maxes out at 6mph, the Sidewalk Operation Mode at 9mph, and the Open Environment Mode reaches 12.5mph.

Segway also offers a small lightweight version, a rugged cross-terrain vehicle, and a golf transport. This is one of the coolest ways to travel!

Secret Agent Walking Stick

If you need a walking stick, this one is perfect. If you don't need help walking, you might still want to use this device. It is much more than an ordinary cane. Like 007 or Maxwell Smart, this device has a few secret functions and even a secret compartment.

The stick itself is lightweight black aluminum with a non-skid rubber tip. It can adjust from 33 inches to 37 inches. Although it's not real wood, the handle looks just like polished burled walnut wood. The grip-shape on the bottom of the handle makes the stick easy to hold.

Insert two AAA batteries, and you will see one of the more useful secrets of this walking stick. Just press the button on the front of the handle, and a flashlight beam on the front of the handle illuminates the way. Press the button again, and the flashlight goes off while the back of the handle flashes a red safety light. Press again, and both are illuminated. This feature is good for anyone at night, even if you're just walking the dog.

> **Model:** Secret Agent Walking Stick; Item # CA015
> **Manufacturer:** Gold Violin
> **URL: www.goldviolin.com**
> **Price: $39.95**

Last, but not least, the Secret Agent Walking Stick has a small hidden slide-out container in the handle that you can use for pills or any other small objects you might want to hide. The stick folds to 12.5 inches for travel and comes with a nylon carrying case. It is also available with a gold-colored shaft.

Vibrating Pillow LCD Alarm Clock

Whether you are taking a business trip or going camping with the kids, this portable alarm clock ensures that you don't oversleep. It's small enough to slip under your pillow, so you can also use it at home when you have to get up early and you don't want to wake your mate. It's also perfect if you have difficulty hearing or you are a heavy sleeper.

Model: Item # 990220
Manufacturer: Reizen
URL: www.maxiaids.com
Price: $24.95

The alarm features the vibrating alarm system and also has dual alarms. The clock displays the day, month, temperature, and time on a monochrome display with a bright green backlight. It works on two AAA batteries and comes with a nice leather case for the times you want to take it on the road.

The Ape Case "Tri-Fold Traveler"

The Ape Case Tri-Fold Traveler is a tri-fold wallet with 14 pockets and compartments. When the case folds, it is held in place with strong Velcro closures. The 5.5" × 6.5" × 3" Traveler has plenty of room for credit cards, passport, digital media cards, money, maps, and more. The front pouch is large enough for a small or mid-sized digital camera, PDA, or other gadget. A cell phone fits in the outside or inside pouch. Inside pockets are

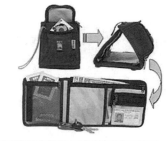

Model: AC252
Manufacturer: Norazza
URL: www.norazza.com
Price: $29.99

zippered for added protection. The bright yellow lining makes it easier to locate items inside the case.

This Ape Case is made of water-resistant, heavy-duty nylon. It is padded to protect equipment and comes with a lifetime warranty. It has a handy, but removable, over-the-shoulder carrying strap and a loop that can attach to your belt. This well-designed case might just become a constant travel companion.

Traveling with a Computer

After you realize how much information is on the Internet and get connected with friends and family through email, Internet access becomes something you just don't want to live without. Using a computer when you travel or when you move to different locations at different times of the year might become a way of life for you. Following are a few gadgets that can help when you need to use your computer on the road.

Mac Mini

The Mini at 6.5" × 6.5" deep × 2" high is one of the tiniest computers. It weighs in at less than 3 pounds, making it lighter than any of its competitors. So why is a computer like this mentioned in the chapter about travel?

Well, the Mac Mini is perfect for snowbirds. Many people spend their summers up North and their winters down South. Snowbirds or anyone who travels to alternative locations will find that this small computer is perfect for them. Transferring of files and changing of email settings are a thing of the past. Just have a monitor, mouse, and keyboard in each location and take the Mac Mini along with you. It even comes in a small box with a handle on top for easy transport. It's also a great choice for RVers.

The Mac Mini is a full-fledged, quite capable computer. It comes with the Apple OS X Tiger operating system and the iLife collection of digital media applications. Under the hood, the Mini sports a 1.25GHz G4 processor, 256MB of RAM, a 40GB hard drive, and a combination CD-burner, DVD-ROM drive. Its ATI's Radeon 9200 graphics chip has 32MB of graphics memory. Ports that you can access

Model: Mac Mini
Manufacturer: Apple
URL: www.apple.com
Price: $499

from the back of the Mini include a FireWire port, two USB ports, an Ethernet port, a modem, and both digital and analog monitor ports. If you buy directly from Apple, you can custom-configure the Mac Mini for your needs.

Canary Digital Hotspotter

You can find wireless hotspots all over the country. These are places where you can use a wireless-enabled laptop or other device to hook up to the Internet. Parks, restaurants, airports, and hotels are just some of

the many places that offer wireless access to the Internet. The only problem is that when you travel, it's hard to locate these hotspots. They are not listed in the Yellow Pages, and you can't rely on asking gas station attendants because many people don't even know these hotspots exist.

This problem is solved with the Canary Digital Hotspotter. This wonderful device detects hotspots where you can access a wireless network. Although it is slightly more expensive, the Canary works better than other hotspot finders I've tried. It also offers more information. The small LDC screen gives you the name of the network, the signal strength, the encryption status, and the channel that the

Model: Digital Hotspotter Model #HS10
Manufacturer: Canary Wireless
URL: www.canarywireless.com
Price: $59.95

signal is transmitting on. It even provides specific information about each network when multiple networks are available.

Because the Hotspotter is only about 2 1/2 inches square and 1 inch thick, it is easy to take it with you, so you will always be able to find wireless Internet service.

Travel Connection Kit—World Pack

Have you ever wound up in a foreign country unable to use your laptop computer because that country has different power or telephone plugs? Well, I have, and I guarantee that it is no fun to be in that situation.

This set of power plug and telephone travel adapters keeps your power devices working and keeps you connected no matter where you travel. This kit contains adapters and connectors for international telephone and power connections. It includes 18 phone adapters and 5 power adapters. The best part is that they come in a zippered case with each adapter visible and clearly labeled.

This set gives you adapters that can be used in more than 100 countries, allowing you to easily overcome travel power plug and telephone connection problems. Other kits with fewer adapters that cost less are also available. If you travel a lot and need to power your equipment or connect your laptop, this kit is a wonderful travel companion.

Model: Travel Connection Kit—World Pack
Manufacturer: Targus
URL: www.targus.com
Price: $99.99

Skooba™ Satchel Laptop Case

When you travel, you probably take along a laptop computer. A laptop case should look good and offer protection and organization for your equipment. The Skooba Laptop Case does all these plus more.

The Skooba can handle a laptop with up to a 17-inch screen. It is an attractive case made of tough ballistic nylon. It comes in black with blue trim, gray/red, olive/red, and

Model: Skooba™ Satchel
Manufacturer: RoadWired
URL: www.roadwired.com
Price: $99

khaki/black. RoadWired also remembered the female computer user with its pink and black Skooba.

Protection is provided by a patented Air Square fabric that surrounds the laptop with hundreds of shock-absorbing air-filled pouches. The Skooba also offers excellent organizational features, with more than 15 pockets to hold files, magazines, keys, laptop accessories, CDs, and other gadgets. It also has a mesh pouch for a water bottle and an easily accessible ID pocket in the flap.

Another great feature is that the case itself is lightweight compared to others. It also has an oversized shoulder pad with the Air Square cushioning that makes a fully loaded case easier to carry. RoadWired has smaller Skooba Sleeves that are designed to protect a laptop while it is in a briefcase or larger bag.

Sandy's Lingo List

The world of technology has created some crazy new words. Here are explanations for a few of the more unusual words used in this chapter.

encryption—A way of translating data into a secret code. Encryption is used to secure data so that only the proper recipients can read it. It is similar to coded messages that were used during wartime, but computer technology today makes computer data much more secure.

global positioning system (GPS)—A navigational system that is formed by 24 satellites orbiting the earth and GPS receivers on or in close proximity to the ground. Digital radio signals beamed from the satellites allow the receiver to determine the exact longitude and latitude of its current location. Some GPS receivers can also calculate the altitude of the receiver.

hotspots—Areas where a wireless broadband network provides Internet service for wireless-enabled devices such as laptop computers. Hotspots are often found in parks, restaurants, hotels, and airports. Some are free to anyone, whereas others require an access fee.

Index

Symbols

A

B

C

E

Rather than having you read through a lot of text, Easy lets you learn visually. Users are introduced to topics of technology, hardware, software, and computersin a friendly, yet motivating, manner.

Easy Digital Cameras
Mark Edward Soper
ISBN: 0-7897-3077-4
$19.99 USA/$28.99 CAN

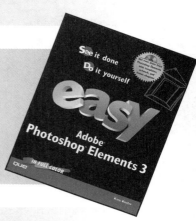

Easy Adobe Photoshop Elements 3
Kate Binder
ISBN: 0-7897-3330-7
$19.99 USA/$28.99 CAN

Easy Microsoft Windows® XP, Home Edition
Third Edition
Shelley O'Hara
ISBN: 0-7897-3337-4
$19.99 USA/$28.99 CAN

Easy Digital Home Movies
Jake Ludington
ISBN: 0-7897-3114-2
$19.99 USA/$28.99 CAN